Introduction to Differential Calculus Systematic Studies with Engineering Applications

Introduction to Differential Calculus Systematic Studies with Engineering Applications

Editor

Jai Rathod

Introduction to Differential Calculus Systematic Studies with Engineering Applications
Edited by **Jai Rathod**

Printed in 2017
ISBN: 978-1-68117-184-5
Library of Congress Control Number: 2015949113

© 2016 by
SCITUS Academics LLC,
616, Corporate Way, Suite 2, 4766,
Valley Cottage, NY 10989

www.scitusacademics.com

Preface

Differential calculus is a subfield of calculus concerned with the study of the rates at which quantities change. It is one of the two traditional divisions of calculus, the other being integral calculus.

In differential calculus, primary objects of study are the derivative of a function, related notions such as the differential, and their applications. The derivative of a function at a chosen input value describes the rate of change of the function near that input value. The process of finding a derivative is called differentiation. Geometrically, the derivative at a point is the slope of the tangent line to the graph of the function at that point, provided that the derivative exists and is defined at that point. For a real-valued function of a single real variable, the derivative of a function at a point generally determines the best linear approximation to the function at that point.

Differential calculus and integral calculus are associated by the fundamental theorem of calculus, which states that differentiation is the reverse process to integration.Differentiation has applications to nearly all quantitative disciplines.Derivatives are frequently used to find the maxima and minima of a function. Equations involving derivatives are called differential equations and are fundamental in describing natural phenomena. Derivatives and their generalizations appear in many fields of mathematics, such as complex analysis, functional analysis, differential geometry, measure theory and abstract algebra.

Introduction to Differential Calculus: Systematic Studies with Engineering Applications for Beginners presents the fundamental theories and methods of differential calculus and shows how the discussed concepts can be applied to real-world problems in engineering and the physical sciences. The book sets a solid foundation before advancing to specific calculus methods, demonstrating the connections between differential calculus theory and its applications.

Contents

Mean Square Numerical Methods for Initial Value Random Differential Equations

Magdy A. El-Tawil[1] and Mohammed A. Sohaly[2]

[1]Engineering Mathematics Department, Faculty of Engineering, Cairo University, Giza, Egypt
[2]Mathematic Departments, Faculty of Science, Mansoura University, Mansoura, Egypt

ABSTRACT

In this paper, the random Euler and random Runge-Kutta of the second order methods are used in solving random differential initial value problems of first order. The conditions of the mean square convergence of the numerical solutions are studied. The statistical properties of the numerical solutions are computed through numerical case studies.

INTRODUCTION

Random differential equations (RDE) are defined as differential equations involving random inputs. In recent years, increasing interest in the numerical solution of (RDE) has led to the progressive development of several numerical methods. This paper is interested in studying the following random differential initial value problem (RIVP) of the form:

$$\frac{dX}{dt} = f(t, X), \ X(t_0) = X_0 \tag{1.1}$$

Randomness may exist in the initial value or in the differential operator or both. In [1,2], the authors discussed the general order conditions and a global convergence proof is given for stochastic Runge-Kutta methods applied to stochastic ordinary differential equations (SODEs) of Stratonovich type. In [3, 4], the authors discussed the random Euler method and the conditions for the mean square convergence of this problem. In [5], the authors considered a very simple adaptive algorithm based on controlling only the drift component of a time step. Platen, E. [6] discussed discrete time strong and weak approximation methods that are suitable for different applications. Other numerical methods are discussed in [7-12].

In this paper the random Euler and random Runge-Kutta of the second order methods are used to obtain an approximate solution for equation (1.1). This paper is organized as follows. In Section 2, some important preliminaries are discussed. In Section 3, the existence and uniqueness of the solution of random differential initial value problem is discussed and the convergence of random Euler and random Runge-Kutta of the second order methods is discussed. In Section 4, the statistical properties for the exact and numerical solutions are studied. Section 5 presents the solution of some numerical examples of first order random differential equations using random Euler and random Runge-Kutta of the second order methods showing the convergence of the numerical solutions to the exact ones (if possible). The general conclusions are presented in the end section.

PRELIMINARIES

Mean Square Calculus [13]

Definition1: Let us consider the properties of a class of real r.v.'s $X_1, X_2, \cdots X_n$ whose second moments, $E\{X_1^2\}, E\{X_2^2\}, \ldots$ are finite. In this case, they are called "second order random variables", (2.r.v's).

Definition 2: The linear vector space of second order random variables with inner product, norm and distance, is called an L_2-space.

A s.p. $\{X(t), t \in T\}$ is called a "second order stochastic process" (2.s.p) if for $t_1, t_2, \ldots t_n$, the r.v's $\{X(t_1), X(t_2), \ldots X(t_n),\}$ are elements of L_2 -space.

A second order s.p. $\{X(t), t \in T\}$ is characterized by $\|X(t)\|^2 = E\{X^2(t)\} < \infty, t \in T$.

The Convergence in Mean Square

A sequence of r.v's $\{X_n\}$ converges in mean square (m.s) to a random variable X if $\lim_{x \to \infty} \|X_n - X\| = 0$ i.e. $X_n \xrightarrow{\text{m.s}}$ or $\lim_{x \to \infty} X_n - X = 0$ where lim is the limit in mean square sense.

Mean-Square Differentiability

The random process $\{X(t)\}$ is mean-square differentiable at t if $\lim_{x \to 0} \dfrac{X_{t+h} - X_t}{h}$ exists, and is denoted by $\lim_{x \to 0} \dfrac{X_{t+h} - X_t}{h} = \dot{X}_t$

RANDOM INITIAL VALUE PROBLEM (RIVP)

Existence and Uniqueness

Let us have the random initial value problem

$$\frac{dX}{dt} = b(t, X), t \in T = [t_o, t], X(t_0) = X_0 \tag{3.1}$$

Where $X(t)$ is second order random process? This equation is equivalent to integral equation

$$X(t) = X_0 + \int_{t0}^{t} b(X(s), s) ds \tag{3.2}$$

Theorem: (3.1.1)

If we have the random initial value problem (3.1) and suppose the right-hand side function b(t, X) is continuous and satisfies a mean square (m.s) Lipschitz condition in its second argument:

$$\left\| b(t,X) - b(t,Y) \right\| \le c \left\| X - Y \right\|$$

(3.3)

Where C is a constant or

$$\left\| b(t,X) - b(t,Y) \right\| \le c(t) \left\| X - Y \right\| \le c \left\| X - Y \right\|$$

(3.4)

Where c (t) is a continuous function {because in every finite interval c(t) ≤ constant}.

Then the solution of equation (3.1) exists and is unique.

The Proof

The existence can be proved by using successive approximations. Let

$$X_t^0 = X_0$$

(3.5)

And for n ≥ 1

$$X_t^n = X_0 + \int_{t0}^{t} b\left(X_s^{n-1}, s \right) ds.$$

(3.6)

For n = 1 we obtain:

$$\left\| X_t^{(1)} - X_t^{(0)} \right\| = \left\| \int_{t0}^{t} b(s, X_0) ds \right\| \le k \cdot \left| t - t_0 \right|$$

Where

$$\left\| b(t,X) \right\| \le k$$

(3.7)

For n > 1 we obtain:

Mean Square Numerical Methods for Initial Value Random

$$\left\| X_t^{(n)} - X_t^{(n-1)} \right\| = \left\| \int_{t0}^{t} b\left(s, X_s^{n-1}\right) - b\left(s, X_s^{n-2}\right) ds \right\|$$

$$\le \int_{t0}^{t} c \cdot \left\| X_s^{(n-1)} - X_s^{(n-2)} \right\| ds \tag{3.8}$$

Successively, we can obtain the following:

$$\left\| X_t^{(n)} - X_t^{(n-1)} \right\| \le \int_{t_0}^{t} c \left\| X_s^{n-1} - X_s^{n-2} \right\| ds$$

$$\le \int_{t_0}^{t} c \int_{t_0}^{t} c \left\| X_s^{n-2} - X_s^{n-3} \right\| ds ds \le c^3 \int_{t_0}^{t} \int_{t_0}^{t} \int_{t_0}^{t} \left\| X_s^{n-3} - X_s^{n-4} \right\| ds ds ds$$

$$\le c^{n-1} \int_{t_0}^{t} \int_{t_0}^{t} \cdots \int_{t_0}^{t} \left\| X_s^1 - X_s^0 \right\| ds \cdots ds \le c^{n-1} \int_{t_0}^{t} \int_{t_0}^{t} \cdots \int_{t_0}^{t} k \left| s - t_0 \right| d \cdots ds.$$

Hence

$$\left\| X^n - X^{n-1} \right\| \le k c^{n-1} \int_{t_0}^{t} \left[\int_{t_0}^{t} \cdots \int_{t_0}^{t} \left[\int_{t_0}^{t} \left| s - t_0 \right| ds \right] ds \right] \cdots ds \left[ds = k c^{n-1} \frac{\left| t - t_0 \right|^n}{n!} \right. \tag{3.9}$$

Since:

$$\sum_{n=1}^{\infty} k.c^{(n-1)} \frac{\left| t - t_0 \right|}{n!} = \frac{k}{c} e^{c \left\| t - t_0 \right\|} \tag{3.10}$$

is convergent for finite t, Hence we can have the following

$$\left\| X_t^1 - X_t^0 \right\| + \left\| X_t^2 - X_t^1 \right\| + \cdots + \left\| X_t^{n-1} - X_t^{n-2} \right\| + \cdots \le \left(\frac{k \left| t - t_0 \right|}{1!} \right) + \left(\frac{k c \left| t - t_0 \right|^2}{2!} \right) + \cdots +$$

$$\left(\frac{k c^{n-2} \left| t - t_0 \right|^{n-1}}{(n-1)!} \right) + \cdots = \sum_{n=1}^{\infty} k.c^{(n-1)} \frac{\left| t - t_0 \right|^n}{n!} \tag{3.11}$$

Accordingly,

$$\left\| X_t^1 - X_t^0 + X_t^2 - X_t^1 + \cdots + X_t^{n-1} - X_t^{n-2} + X_t^n - X_t^{n-1} \right\| = \lim_{n \to \infty} \left\| X_t^n - X_t^0 \right\| = \left\| X_t \right\|$$

Hence:

$$\left\| X_t^1 - X_t^0 + X_t^2 - X_t^1 + \cdots X_t^{n-1} - X_t^{n-2} + X_t^n - X_t^{n-1} \right\| \leq \left\| X_t^1 - X_t^0 \right\| + \left\| X_t^2 - X_t^1 \right\| \cdot$$

$$+ \cdots + \left\| X_t^{n-1} - X_t^{n-2} \right\| + \left\| X^n - X^{n-1} \right\| = \frac{k}{c} e^{c\|t - t_0\|}$$

This yield $\lim\limits_{x \to \infty} \left\| X_t^n - X_t^0 \right\| = \left\| X_t \right\| \leq \dfrac{k}{c} e^{c\|t - t_0\|}$

Then $\lim\limits_{x \to \infty} X_n$ exists i.e

$$X_t = \lim_{h \to \infty} X_t^n \tag{3.12}$$

Since X_t^n is the general solution of equation (3.6) and X_t is the general solution of equation (3.2).

To prove the uniqueness of the solution, let X_t is a solution of the initial-value problem (3.1), or, which is the same, of the integral equation (3.2), and Y_t is the solution of

$$\frac{dy}{dt} = b\big(Y(t), t\big), t \in T = [t_0, t], \ Y(t_0) = Y_0 \tag{3.13}$$

to prove the uniqueness of the solution we want to prove that

$$X_t = Y_t \tag{3.14}$$

By subtraction (3.2) and the corresponding integral equation for Y_t

$$X_t - Y_t = X_0 - Y_0 + \int_{t_0}^{t} \big(b(X_s, s) - b(Y_s, s)\big)ds \tag{3.15}$$

Since $X_0 = Y_0$ then:

$$\left\| X_t - Y_t \right\| \leq \int_{t_0}^{t} c \cdot \left\| X_s - Y_s \right\| ds \tag{3.16}$$

i.e; $U^t \leq \int_{t_0}^{t} c.U_s ds \tag{3.17}$

Where $U_t \|X_t - Y_t\|$ (3.18)

From equation (3.17) we have:

$$\|U_t\| \le c|t - t_0|\|U_t\|$$ (3.19)

Note that: at $t = t_0$ we obtain $\|U_0\| \le 0$ then: $\|U_0\| = 0$

From (3.19) C must satisfy the following condition:

$$c \ge \frac{1}{|t - t_0|}$$ (3.20)

Which is in contradiction with being an independent free constant, hence the only solution of the integral equation (3.17) is

$$U^t = 0$$ (3.21)

Hence $X_t = Y_t$ i.e., the solution of equation (3.1) exists and is unique.

The Convergence of Euler Scheme for Random Differential Equations in (M.S.) Sense

Let us have the random differential equation

$$\dot{X}(t) = f(X(t), t), t \in T = [t_0, t_1], X(t_0) = X_0$$ (3.22)

Where X_0 is a random variable and the unknown $X(t)$ as well as the right-hand side f (X, t) are stochastic processes defined on the same probability space.

Definitions [6, 7]

- Let g: $T \to L_2$ is an m.s. bounded function and let h > 0 then The "m.s. modulus of continuity of g" is the function

$$W(g, h) = \sup_{|t - t^*| \le h} \|g(t) - g(t^*)\|, \ t, t^* \in T$$

- The function g is said to be m.s uniformly continuous in T if:

$$\lim_{h \to 0} W(g,h) = 0$$

Note that:

(The limit depends on h because g is defined at every t so we can write $W(g,h)=W(h)$)

In the problem (3.22), we find that the convergence of this problem depends on the right hand side (i.e. $f(X(t),t)$ then we want to apply the previous definition on $f(X(t),t)$ hence:

Let $f(X(t),t)$ be defined on $S \times T$ where S is bounded set in L_2

Then we say that f is "randomly bounded uniformly continuous" in S, if

$$\lim_{x \to \infty}(f(x,.),h) = 0$$

(Note that $w\big(f\big(x(.),h\big)\big) = w\big(h\big)$)

Random Mean Value Theorem for Stochastic Processes

The aim of this section is to establish a relationship between the increment $X(t) - X(t_0)$ of a 2-s.p. and its m.s. derivative $\dot{X}(\xi)$ for some ξ lying in the interval $\big[t_0,t\big]$ for $t > t_0$. This result will be used in the next section to prove the convergence of the random Euler method.

Lemma: (3.3.2) [6, 7]

Let Y (t) is a 2-s.p., m.s. continuous on interval $T = \big[t_0,t\big]$. Then, there exists $\xi \in \big[t_0,t\big]$ such that

$$\int_{t_0}^{t} Y(s)ds = Y(\xi)(t - t_0), t_0 < t_1 < t \tag{3.25}$$

The Proof

Since Y(t) is m.s. continuous, the integral process

$\int_{t_0}^{t} Y(ds)$ Is well defined and the correlation function

$\Gamma_y(r,s)$ Is well defined, is a deterministic continuous function on $T \times T$.

For each fixed r, the function $\Gamma_y(r,s)$ is continuous and by the classic mean value theorem for integrals, it follows that:

$$\int_{t_0}^{t} \Gamma_y(r,s)\,ds = \Gamma_y(r,\xi)(t-t_0)$$

$$\forall \xi \in [t_0,t] \tag{3.26}$$

Note that by definition of $\Gamma_y(r,s)$ expression (3.26) can be written in the form

$$\int_{t_0}^{t} E\big[y(r)y(s)\big]\,ds = E\big[y(r)y(\xi)\big](t-t_0)$$

Since $\Gamma_y(r,s) = E\big[y(r)y(s)\big]$

We must prove that for the value ξ satisfying (3.26) one get:

$$\left\| \int_{t_0}^{t} y(s)\,ds - y(\xi)(t-t_0) \right\|^2$$

$$= E\left[\left(\int_{t_0}^{t} y(s)\,ds - y(\xi)(t-t_0) \right)^2 \right] = 0 \tag{3.27}$$

The Proof of (3.27)
As

$$E\left[\left(\int_{t_0}^{t} y(s)\,ds - y(\xi)(t-t_0) \right)^2 \right]$$

$$= E\left[\left(\int_{t_0}^{t} y(s)\,ds \right)^2 \right] - 2E\left[\left(\int_{t_0}^{t} y(s)\,ds \right) y(\xi) \right](t-t_0)$$

$$+ E\left[y(\xi)^2 (t-t_0)^2 \right] \tag{3.28}$$

and since:

$$E\left[\left(\int_{t_0}^{t} y(s)\,ds\right)^2\right] = \int_{t_0}^{t}\int_{t_0}^{t} E\left[y(s)y(r)\right]drds$$

Then by substituting in (3.28)

$$E\left[\left(\int_{t_0}^{t} y(s)\,ds - y(\xi)(t-t_0)\right)^2\right]$$

$$= \int_{t_0}^{t}\int_{t_0}^{t} E\left[y(s)y(r)\right]drds - \int_{t_0}^{t} E\left[y(s)y(\xi)\,ds\right)(t-t_0)$$

$$- \int_{t_0}^{t} E\left[y(s)y(\xi)\,ds\right)(t-t_0) + E\left[y(\xi)y(\xi)(t-t_0)^2\right] \qquad (3.29)$$

And since:

$$\int_{t_0}^{t} E\left[y(r)y(s)\right]ds = E\left[y(r)y(\xi)\right](t-t_0)$$

Then by substituting in (3.28) we have:

i.e. $\left\|\int_{t_0}^{t} y(s)\,ds - y(\xi)(t-t_0)\right\|^2 = 0$

We obtain

$$\int_{t_0}^{t} y(s)\,ds = y(\xi)(t-t_0)$$

Theorem: (3.3.1) [6, 7]

Let $X(s)$ be a m.s. differentiable 2-s.p. in $\left[t_0,t_1\right]$ and m.s. continuous in $T = \left[t_0,t\right]$. Then, there exists $\xi \in \left[t_0,t_1\right]$ such that $X(t) - X(t_0) = \dot{X}(\xi)(t-t_0)$, $t_0 < t_1 < t$

The Proof
The result is a direct consequence of Lemma (3.3.2) applied to the 2-s.p. $Y(t) = \dot{X}(t)$

$$\int_{t0}^{t} \dot{X}(s)ds = \dot{X}(\xi)(t-t_0)$$

And the integral formula

$$\int_{t0}^{t} \dot{X}(s)ds = X(t) - X(t_0) \tag{3.30}$$

The Proof of (3.30)

Let X(t) be a m.s. differentiable on T and let the ordinary function f (t,s) be continuous on $T \times T$ whose partial derivative $\dfrac{\partial f(t,s)}{\partial s}$ exist. If

$$Y(t) = \int_a^t f(t,s)\dot{X}(s)ds \tag{3.31}$$

Then

$$Y(t) = f(t,s)X(s)\Big|_a^t - \int_a^t \frac{\partial f(t,s)}{\partial s} X(s)ds \tag{3.32}$$

Let $f(t,s) = 1$ in Equations (3.31) and (3.32) we have the useful result that:

If $\dot{X}(t)$ is m.s. Riemann integrable on T then:

$$\int_a^t \dot{X}(s)ds = X(t) - X(a), [a,t] \subset T$$

Then we have:

$$X(t) - X(t_0) = \dot{X}(\xi)(t-t_0)$$

The Convergence of Random Euler Scheme

In this section we are interested in the mean square convergence, in the fixed station sense, of the random Euler method defined by

$$X_{n+1} = X_n hf(X_n, t_n), X(t_0) = X_0, n \geq 0 \tag{3.33}$$

Where X_n and $f(X_n, t_n)$ are 2-r.v.'s $h = t_n - t_{n-1}, t_n = t_0 + nh$ and f: $S \times T \rightarrow L_2, S \subset L_2$ satisfies the following conditions:

C1: $f(\underset{\cdot}{X}t)$ is randomly bounded uniformly continuous C2: $f(\underset{\cdot}{X}t)$ satisfies the m.s. Lipschitz condition

$$\|f(x,t)-f(y,t)\| \leq k(t)\|x-y\|$$

Where

$$\int_{t_0}^{t_1} k(t) \leq \infty \qquad\qquad (3.34)$$

Note that under hypothesis C1 and C2, we are interested in the m.s. convergence to zero of the error

$$e_n = X_n - X(t) \qquad\qquad (3.35)$$

Where $X(t)$ is the theoretical solution 2-s.p. of the problem (3.22), $t = t_n = t_0 nh$.

Taking into account (3.22), and Theorem (3.3.1), one gets since from (3.22) we have at $X = t_\xi$ then

$$\dot{X}(t_\xi) = f\left(X(t_\xi), t_\xi\right)$$

Note $\xi \in [t_0, t_1]$ and we can use ξ instead of t_ξ and from Theorem (3.3.1) at $t = t_\xi$ then we have:

$$X(t_\xi) - X(t_0) = \dot{X}(t)(t_\xi - t_0)$$

Then

$$X(t_\xi) - X(t_0) = f\left(X(t_\xi, t_\xi)\right)(t_\xi - t_0)$$

Note that we deal with the interval $(t_n, t_{n+1}) \ni t_\xi \in (t_n, t_{n+1})$ and hence t_0 was the starting in the problem (3.22) and here t_n is the starting and since Euler method deal with solution depend on previous solution and if we have $X(t_n)$ instead of $X(t_0)$ then we can use $X(t_{n+1})$ instead of $X(t_\xi)$.

Then the final form of the problem (3.22) is

$$X(t_{n+1}) = X(t_n) + hf(X(t_\xi, t_\xi)),$$

For some

$$t_\xi \in (t_n, t_{n+1}) \tag{3.36}$$

Now we have the solution of problem (3.22) is $X(t_n)$

At $t = t_n$ then $X(t_n) = X(t)$ and the solution of Euler method (3.33) is X_n

Then we can define the error

$$e_n = X_n - X(t_n)$$

$$e_n = X_n - X(t)$$

By (3.33) and (3.36) it follows that

$$X_{n+1} - X(t_{n+1})$$
$$= X_n + hf(X_n, t_n) - X(t_n) - hf(X(t_\xi, t_\xi))$$

This implies

$$e_{n+1} = X_n - X(t_n) + h\{f(X_n, t_n) - f(X(t_\xi), t_\xi)\}$$

Hence

$$\|e_{n+1}\| = \|X_n - X(t_n) + h\{f(X_n, t_n) - f(X(t_\xi), t_\xi)\}\|$$
$$\leq \|X_n - X(t_n)\| + h\|f(X(t_\xi), t_\xi) - f(X_n, t_n)\| \tag{3.37}$$

Since:

$$\left\| f\left(X\left(t_\xi\right),t_\xi\right)-f\left(X_n,t_n\right)\right\|$$

$$=\left\| f\left(X\left(t_\xi\right),t_\xi\right)-f\left(X\left(t_\xi\right),t_n\right)+f\left(X\left(t_\xi\right),t_n\right)\right.$$

$$\left.+f\left(X\left(t_n\right),t_n\right)-f\left(X\left(t_n\right),t_n\right)-f\left(X_n,t_n\right)\right\|$$

$$\leq\left\| f(X(t_\xi),t_\xi)-f(X(t_\xi),t_n)\right\|$$

$$+\left\| f(X(t_\xi),t_n)-f(X(t_n),t_n)\right\|$$

$$+\left\| f(X(t_n),t_n)-f(X_n,t_n)\right\|$$

(3.38)

Since the theoretical solution X is m.s. bounded in

$[t_0,t_1]$, $\sup\limits_{t_0 \leq t \leq t_1}\left\|X(t)\right\|\leq M<\infty$ and Under hypothesis C1, C2 We obtain

- $\left\| f\left(X(t_\xi),t_\xi\right)-f\left(X(t_\xi),t_n\right)\right\|=w(h)$

- $\left\| f\left(X(t_\xi),t_\xi\right)-f\left(X(t_\xi),t_n\right)\right\|\leq k(t_n)Mh(^*)$

Since $k(t_n)$ is Lipschitz constant (from C2) and from Theorem (3.3.1) we have $X(X_n)-X(t_0)=\dot{X}(\xi)(t-t_0)$ and note that the two points are $X(X_\xi)$ and $X(X_n)$ in (*) then we have:

$$\left\| X\left(t_\xi\right)-X\left(t_n\right)\right\|=\left\|\dot{X}\left(\xi\right)\right\|\left| t_\xi -t_n\right|\leq Mh$$

Since $\left| t_\xi -t_n\right|=h$ and $M=\sup\left\|\dot{X}(t)\right\|$

- $\left\| f\left(X\left(t_n\right),t_n\right)-f\left(X_n,t_n\right)\right\|$

  $\leq k\left(t_n\right)\left\| X\left(t_n\right)-X_n\right\|=k\left(t_n\right)\left\| e_n\right\|$

Then by substituting in (3.38) we have

$$\left\| f\left(X\left(t_\xi\right),t_\xi\right)-f\left(X_n,t_n\right)\right\|\leq w(h)+k\left(t_n\right)Mh+k\left(t_n\right)\left\| e_n\right\|$$

(3.39)

Then by substituting in (3.37) we have

Mean Square Numerical Methods for Initial Value Random

$$\left\|e_{n+1}\right\| \le \left\|e_n\right\| + h\left[w(h) + k(t_n)Mh + k(t_n)\left\|e_n\right\|\right] = \left(1 + k(t_n)h\right)\left\|e_n\right\| + h\left[w(h) + k(t_n)Mh\right]$$

$$\le \left(1 + K(t_n)h\right)\left[\left(1 + k(t_n)h\right)\left\|e_{n-1}\right\| + h\left[w(h) + k(t_n)Mh\right]\right] + h\left[w(h) + k(t_n)Mh\right]$$

$$= \left(1 + K(t_n)h\right)^2 \left\|e_{n-1}\right\| + h\left[w(h) + k(t_n)Mh\right]\left[1 + \left(1 + k(t_n)h\right)\right]$$

$$\le \left(1 + K(t_n)h\right)^3 \left\|e_{n-2}\right\| + h\left[w(h) + k(t_n)Mh\right]\left[1 + \left(1 + k(t_n)h\right) + \left(1 + k(t_n)h\right)^2\right]$$

$$< \left(1 + K(t_n)h\right)^{n+1} \left\|e_0\right\| + h\left[w(h) + k(t_n)Mh\right]\left[1 + \left(1 + k(t_n)h\right) + \left(1 + k(t_n)h\right)^2 + \cdots + \left(1 + k(t_n)h\right)^n\right]$$

Since:

$$\left[1 + \left(1 + k(t_n)h\right) + \left(1 + k(t_n)h\right)^2 + \cdots + \left(1 + k(t_n)h\right)^n\right]$$

Is geometrical sequence

Then:

$$\left[1 + \left(1 + k(t_n)h\right) + \left(1 + k(t_n)h\right)^2 + \cdots + \left(1 + k(t_n)h\right)^n\right]$$

$$= \frac{\left(1 + k(t_n)h\right)^n - 1}{k(t_n)h}$$

Then we get

$$\left\|e_{n+1}\right\| \le \left(1 + K(t_n)h\right)^{n+1}\left\|e_0\right\|$$

$$+ \left[w(h) + k(t_n)Mh\right]\frac{\left[\left(1 + k(t_n)h\right)^n - 1\right]}{k(t_n)}$$

Taking into account that $e_0 = 0$ where

$$e_0 = X_0 - X(t_0) = 0.$$

$$\|e_{n+1}\| \le \left[w(h) + k(t_n)Mh\right]\frac{\left[(1+k(t_n)h)^n - 1\right]}{k(t_n)}$$

$$\lim_{h\to 0}\|e_{n+1}\| \le \lim_{h\to 0}\left[w(h) + k(t_n)Mh\right]\frac{\left[(1+k(t_n)h)^n - 1\right]}{k(t_n)}$$

$$= \lim_{h\to 0}\frac{\left[w(h) + k(t_n)Mh\right]}{k(t_n)}\left[(1+K(t_n)h)^n\right]$$

$$- \lim_{h\to 0}\frac{\left[w(h) + k(t_n)Mh\right]}{k(t_n)} \tag{3.40}$$

Note that:

The term: $\displaystyle\lim_{h\to 0}\frac{[w(h) + k(t_n)Mh]}{k(t_n)} = 0$ as

$$h \to 0\left(w(h) \to 0 \ni \text{ as } h \to 0\right) \tag{3.41}$$

And the second term:

$$\lim_{h\to 0}\frac{[w(h) + k(t_n)Mh]}{k(t_n)}\left[(1+K(t_n)h)^n\right]$$

We have:

$$\lim_{h\to 0}\frac{\left[w(h) + k(t_n)Mh\right]}{k(t_n)}[(1+K(t_n)h)^n]$$

$$= \lim_{h\to 0}\frac{\left[w(h) + k(t_n)Mh\right]}{k(t_n)}\lim_{h\to 0}\left[(1+K(t_n)h)^n\right] \tag{3.42}$$

The first limit in (3.42) equal zero and:

The computation of $\displaystyle\lim_{h\to 0}\left[(1+K(t_n)h)^n\right]$ as follows:

Let $y = \displaystyle\lim_{h\to 0}\left[(1+K(t_n)h)^n\right]$ then by tacking the logarithm of the two sides we have:

$$\ln y = \ln \lim_{h \to 0} \left[\left(1 + K(t_n)h\right)^n \right]$$

$$= \lim_{h \to 0} \ln \left[\left(1 + K(t_n)h\right)^n \right]$$

$$= \lim_{h \to 0} n \ln \left[\left(1 + K(t_n)h\right) \right]$$

$$= \lim_{h \to 0} \frac{t_n - t_0}{h} \ln \left[\left(1 + K(t_n)h\right) \right]$$

$$= \lim_{h \to 0} \frac{t_n - t_0}{h} \ln \left[\left(1 + K(t_n)h\right) \right]$$

$$= \lim_{h \to 0} \frac{(t_n - t_0)\left[\ln\left(1 + k(t_n)h\right) \right]}{h}$$

By using the (L'Hospital's Rule):

$$\lim_{h \to 0} \frac{(t_n - t_0)\left[\ln\left(1 + k(t_n)h\right) \right]}{h}$$

$$= \lim_{h \to 0} \frac{(t_n - t_0)\dfrac{1}{(1 + k(t_n)h)}k(t_n)}{1}$$

$$= \lim_{h \to 0} \frac{(t_n - t_0)k(t_n)}{1 + k(t_n)h} = (t - t_0)k(t)$$

(3.43)

Then $\ln y = (t - t_0)k(t_n)$ which implies that $y = e^{(t-t_0)k(t_n)}$ hence

$$\lim_{h \to 0} \left[\left(1 + K(t_n)h\right)^n \right] = e^{(t-t_0)k(t_n)}$$

By substituting in (3.42):

$$\lim_{h \to 0} \frac{\left[w(h) + k(t_n)Mh \right]}{k(t_n)} \left[\left(1 + K(t_n)h\right)^n \right]$$

$$= 0 \times e^{(t-t_0)k(t_n)} = 0$$

(3.44)

By substituting from (3.44) and (3.42) in (3.40) hence $\lim\limits_{h\to 0}\|e_{n+1}\|\to 0$

i.e., $\{e_n\}$ converge in m.s to zero as $h\to 0$ hence $X_n \xrightarrow{\text{m.s}} X(t_n) = X(t)$.

The Convergence of Runge-Kutta of Second Order Scheme for Random Differential Equations in Mean Square Sense

In this section we are interested in the mean square convergence, in the fixed station sense, of the random Runge-Kutta of second order method defined by

$$X_{n+1} = X_n + \frac{h}{2}[f(X_n,t_n) + f(X_n + f(X_n,t_n),t_{n+1})],$$

$$X(t_0) = X_0, \ n \geq 0 \tag{3.45}$$

Where X_n and $f(X_n,t_n)$ are 2-r.v.'s, $h = t_n - t_{n-1}, t_n = t_0 + nh$ and $f:$, $S \times T \to L_2, S \subset L_2$ satisfies the following conditions:

C1: $f(X,t)$ is randomly bounded uniformly continuous C2: $f(X,t)$ satisfies the m.s. Lipschitz condition $\|f(x,t) - f(y,t)\| \leq k(t)\|x - y\|$ where

$$\int_{t_0}^{t_1} k(t) \leq \infty \tag{3.46}$$

Note that under hypothesis C1 and C2, we are interested in the m.s. convergence to zero of the error

$$e_n X_n - X(t) \tag{3.47}$$

Where $X(t)$ is the theoretical solution 2-s.p. of the problem (3.22), $t_n = t_0 + nh$.

Taking into account (3.22), and Theorem (3.3.1), one gets since from (3.22) we have at $t = t_\xi$ then

$$\dot{X}(t_\xi) = f\left(x(t_\xi),t_\xi\right)$$

Note $\xi \in [t_0,t_1]$ and we can use ξ instead of t_ξ

And from Theorem (3.3.1) at $t = t_\xi$ then we obtain

$$X(t_\xi) - X(t_0) = \dot{X}(t)(t_\xi t_0) \Rightarrow$$

$$X(t_\xi) - X(t_0) = f\left(X(t_\xi, t_\xi)(t_\xi - t_0)\right)$$

Note that we deal with the interval $(t_n, t_{n+1}) \ni t_\xi \in (t_n, t_{n+1})$ and hence t_0 was the starting in the problem (3.22) and here t_n is the starting and since Euler method deal with solution depend on previous solution and if we have $X(t_n)$ instead of $\Rightarrow X(t_0)$ we can use $X(t_{n+1})$ instead of $X(t_\xi)$ then the final form of the problem (3.22) is

$$X(t_{n+1}) = X(t_n) + hf\left(X(t_\xi, t_\xi)\right), \text{ for some } t_\xi \in (t_n, t_{n+1}) \tag{3.48}$$

Now we have the solution of problem (3.22) is $X(t_n)$

At $t = t_n$ then $X(X_n) = X(t)$ and the solution of Runge-Kutta of 2 order method (3.45) is X_n

Then we can define the error $e_n = X_n - X(t)$

By (3.45) and (3.48) it follows that

$$X_{n+1} - X(t_{n+1})$$

$$= X_n + \frac{h}{2}\left[f(X_n, t_n) + f\left(X_n + f(X_n, t_n), t_{n+1}\right)\right] - X(t_n)$$

$$- \frac{h}{2}f\left(X(t_\xi), t_\xi\right) - \frac{h}{2}f\left(X(t_\xi), t_\xi\right)$$

Then we obtain:

$$e_{n+1} = X_n - X(t_n) + \frac{h}{2}\left\{f(X_n, t_n) - f\left(X(t_\xi), t_\xi\right)\right\}$$

$$+ \frac{h}{2}\left\{f\left(X_n + f(X_n, t_n), t_{n+1}\right) - f\left(X(t_\xi), t_\xi\right)\right\}$$

By taking the norm for the two sides:

$$\|e_{n+1}\| = \left\| X_n - X(t_n) + \frac{h}{2}\{f(X_n,t_n) - f(X(t_\xi),t_\xi)\} \right.$$

$$\left. + \frac{h}{2}\{f(X_n + f(X_n,t_n),t_{n+1}) - f(X(t_\xi),t_\xi)\} \right\|$$

$$\leq \|X_n - X(t_n)\| + \frac{h}{2}\|f(X(t_\xi),t_\xi) - f(X_n,t_n)\|$$

$$+ \frac{h}{2}\|f(X_n + f(X_n,t_n),t_{n+1}) - f(X(t_\xi),t_\xi)\|$$

(3.49)

Since:

$$\|f(X(t_\xi),t_\xi) - f(X_n,t_n)\|$$

$$= \|f(X(t_\xi),t_\xi) - f(X(t_\xi),t_n) + f(X(t_\xi),t_n)$$

$$+ f(X(t_n),t_n) - f(X(t_n),t_n) - f(X_n,t_n)\|$$

$$\leq \|f(X(t_\xi),t_\xi) - f(X(t_\xi),t_n)\| + \|f(X(t_\xi),t_n) - f(X(t_n),t_n)\|$$

$$+ \|f(X(t_n),t_n) - f(X_n,t_n)\|$$

(3.50)

Since the theoretical solution X is m.s. bounded in

$$[t_0,t_1], \quad \sup_{t_0 \leq t \leq t_1}\left\|\dot{X}(t)\right\| \leq M < \infty \text{ and under hypothesis C1, C2 we have}$$

- $\left\|f(X(t_\xi),t_\xi) - f(X(t_\xi),t_n)\right\| = w(h)$

- $\left\|f(X(t_\xi),t_{n-1}) - f(X(t_n),t_{n+1})\right\| \leq k(t)Mh(^*)$

Where $k(t_n)$ Is Lipschitz constant (from C2) and:

From Theorem (3.3.1) we have

$X(t) - X(t_0) = \dot{X}(\xi)(t - t_0)$ And note that the two points Are $k(t_\xi)$ and $k(t_n)$ in (*) then

$$\left\|X(t_\xi) - X(t_n)\right\| = \left\|\dot{X}(\xi)\right\|\left|t_\xi - t_n\right| \leq Mh$$

Where $\left|t_\xi - t_n\right| = h$ and $M = \sup\limits_{t_0 \le t \le t_1}\left\|\dot{X}(t)\right\|$

- $$\left\|f\left(X(t_n),t_n\right)-f\left(X_n,t_n\right)\right\|$$
$$\le k(t_n)\left\|X(t_n)-X_n\right\|=k(t_n)\left\|e_n\right\|$$

Then by substituting in (3.50) we have

$$\left\|f\left(X(t_\xi),t_\xi\right)-f\left(X_n,t_n\right)\right\|\le w(h)+k(t_n)Mh+k(t_n)\left\|e_n\right\|$$

(3.51)

And another term:

$$\left\|f\left(X_n+f(X_n,t_n),t_{n+1}\right)-f\left(X(t_\xi),t_\xi\right)\right\|$$
$$=\left\|f\left(X(t_\xi),t_\xi\right)-f\left(X_n+f(X_n,t_n),t_{n+1}\right)\right\|$$
$$=\left\|f\left(X(t_\xi),t_\xi\right)-f\left(X(t_\xi),t_{n+1}\right)+f\left(X(t_\xi),t_{n+1}\right)\right.$$
$$\left.+f\left(X(t_n),t_{n+1}\right)-f\left(X(t_n),t_{n+1}\right)-f\left(X_n+f(X_n,t_n),t_{n+1}\right)\right\|$$
$$\le\left\|f\left(X(t_\xi),t_\xi\right)-f\left(X(t_\xi),t_{n+1}\right)\right\|+\left\|f\left(X(t_\xi),t_{n+1}\right)-f\left(X(t_n),t_{n+1}\right)\right\|$$
$$+\left\|f\left(X(t_n),t_{n+1}\right)-f\left(X_n+f(X_n,t_n),t_{n+1}\right)\right\|$$
$$\le w(h)+k(t_{n+1})Mh+k(t_{n+1})\left[\left\|e_n\right\|-M\right]$$

Since:

- $\left\|f\left(X(t_\xi),t_\xi\right)-f\left(X(t_\xi),t_n\right)\right\|\le w(h)$
- $\left\|f\left(X(t_\xi),t_{n-1}\right)-f\left(X(t_n),t_{n+1}\right)\right\|\le k(t_{n+1})Mh$

Where $k(t_n)$. Is Lipschitz constant (from C2) and:

From Theorem (3.3.1) we have

$X(t)-X(t_0)=\dot{X}(\xi)(t-t_0)$ And note that the two points are $X(X_\xi)$ and $X(t_n)$ in (*) then we have:

$$\left\|X(t_\xi)-X(t_n)\right\|=\left\|\dot{X}(\xi)\right\|\left|t_\xi-t_n\right|\le Mh$$

Where $\left|t_\xi-t_n\right|=h$ and $M=\sup\limits_{t_0\le t\le t_1}\left\|\dot{X}(t)\right\|$

And the last term:

$$\left\| f\left(X\left(t_n\right),t_{n+1}\right)-f\left(X_n+f\left(X_n,t_n\right).t_{n+1}\right)\right\|$$

$$\leq k\left(t_{n+1}\right)\left\|X\left(t_n\right)-X_n-f\left(X_n,t_n\right)\right\|$$

$$\leq k\left(t_{n+1}\right)\left\|X\left(t_n\right)-X_n\right\|-\left\|f\left(X_n,t_n\right)\right\|$$

$$= k\left(t_{n+1}\right)\left[\left\|e_n\right\|-M\right]$$

Then by substituting in (3.49) we have

$$\left\|e_{n+1}\right\|\leq\left\|e_n\right\|+\frac{h}{2}\left[w(h)+k(t_n)Mh+k(t_n)\left\|e_n\right\|\right]+\frac{h}{2}\left[w(h)+k(t_{n+1})Mh+k(t_{n+1})\left[\left\|e_n\right\|-M\right]\right]$$

$$=\left\|e_n\right\|\left(1+\frac{h}{2}k(t_n)+k(t_{n+1})\right)+\frac{h}{2}\left[2w(h)+hk(t_n)M+hk(t_{n+1})M-K(t_{n+1})M\right]$$

$$\leq\left\{\left\|e_{n-1}\right\|\left(1+\frac{h}{2}k(t_n)+k(t_{n+1})\right)+\frac{h}{2}\left[2w(h)+hk(t_n)M+hk(t_{n+1})M-K(t_n)M\right]\right\}$$

$$\left\{\left(1+\frac{h}{2}k(t_n)+k(t_{n+1})\right)\right\}+\frac{h}{2}\left[2w(h)+hk(t_n)M+hk(t_{n+1})M-K(t_{n+1})M\right]$$

$$=\left\|e_{n-1}\right\|\left(1+\frac{h}{2}k(t_n)+k(t_{n+1})\right)^2+\frac{h}{2}\left[2w(h)+hk(t_n)M+hk(t_{n+1})M-K(t_{n+1})M\right]\left(2+\frac{h}{2}k(t_n)+k(t_{n+1})\right)$$

$$\leq\left\{\left\|e_{n-2}\right\|\left(1+\frac{h}{2}k(t_n)+k(t_{n+1})\right)+\frac{h}{2}\left[2w(h)+hk(t_n)M+hk(t_{n+1})M-K(t_n)M\right]\right\}$$

$$\left\{\left(1+\frac{h}{2}k(t_n)+k(t_{n+1})\right)^2\right\}+\frac{h}{2}\left[2w(h)+hk(t_n)M+hk(t_{n+1})M-K(t_{n+1})M\right]\left(2+\frac{h}{2}k(t_n)+k(t_{n+1})\right)$$

$$=\left\|e_{n-2}\right\|\left(1+\frac{h}{2}k(t_n)+k(t_{n+1})\right)^3+\frac{h}{2}\left[2w(h)+hk(t_n)M+hk(t_{n+1})M-K(t_{n+1})M\right]$$

$$\left[1+\left(1+\frac{h}{2}k(t_n)+k(t_{n+1})\right)+\left(1+\frac{h}{2}k(t_n)+k(t_{n+1})\right)^2\right]$$

Then we have:

$$\left\|e_{n+1}\right\|\leq\left\|e_0\right\|\left(1+\frac{h}{2}k(t_n)+k(t_{n+1})\right)^{n+1}+\frac{h}{2}\left[2w(h)+hk(t_n)M+hk(t_{n+1})M-K(t_{n+1})M\right]$$

$$\left[1+\left(1+\frac{h}{2}k(t_n)+k(t_{n+1})\right)+\left(1+\frac{h}{2}k(t_n)+k(t_{n+1})\right)^2+\cdots+\left(1+\frac{h}{2}k(t_n)+k(t_{n+1})\right)^n\right]$$

Since:

$$\left[1+\left(1+\frac{h}{2}k(t_n)+k(t_{n+1})\right)+\left(1+\frac{h}{2}k(t_n)+k(t_{n+1})\right)^2+\cdots+\left(1+\frac{h}{2}k(t_n)+k(t_{n+1})\right)^n\right]$$

is geometrical sequence then we have:

$$[1+\left(1+\frac{h}{2}k(t_n)+k(t_{n+1})\right)+\left(1+\frac{h}{2}k(t_n)+k(t_{n+1})\right)^2+\cdots+\left(1+\frac{h}{2}k(t_n)+k(t_{n+1})\right)^n]=\frac{\left[\left(1+\frac{h}{2}k(t_n)+k(t_{n+1})\right)^n\right]-1}{\frac{h}{2}k(t_n)+k(t_{n+1})}$$

Then we get:

$$\|e_{n+1}\|\leq\|e_0\|\left(1+\frac{h}{2}k(t_n)+k(t_{n+1})\right)^{n+1}+\frac{h}{2}\left[2w(h)+hk(t_n)M+hk(t_{n+1})M-K(t_{n+1})M\right]\frac{\left[\left(1+\frac{h}{2}k(t_n)+k(t_{n+1})\right)^n\right]-1}{\frac{h}{2}k(t_n)+k(t_{n+1})}$$

Taking into account that $e_0=0$ where

$$e_0=X_0-X(t_0)=0$$

$$\|e_{n+1}\|\leq\frac{h}{2}\left[2w(h)+hk(t_n)M+hk(t_{n+1})M-k(t_{n+1})M\right]\frac{\left[\left(1+\frac{h}{2}k(t_n)+k(t_{n+1})\right)^n\right]-1}{\frac{h}{2}k(t_n)+k(t_{n+1})}$$

$$\lim_{h\to0}\|e_{n+1}\|\leq\lim_{h\to0}\frac{h}{2}\left[2w(h)+k(t_n)Mh+hk(t_{n+1})M-K(t_{n+1})M\right]\frac{\left[\left(1+\frac{h}{2}k(t_n)+k(t_{n+1})\right)^n\right]-1}{\frac{h}{2}k(t_n)+k(t_{n+1})}$$

$$=\lim_{h\to0}\frac{\frac{h}{2}\left[2w(h)+k(t_n)Mh+k(t_{n+1})Mh-k(t_{n+1})M\right]}{\frac{h}{2}k(t_n)+k(t_{n+1})}\left[\left(1+\frac{h}{2}K(t_n)+k(t_{n+1})\right)^n\right]$$

$$-\lim_{h\to0}\frac{\frac{h}{2}\left[2w(h)+k(t_n)Mh+k(t_{n+1})Mh-k(t_{n+1})M\right]}{\frac{h}{2}k(t_n)+K(t_{n+1})}$$

$$(3.52)$$

Note that:

The term:

$$\lim_{h \to 0} \frac{\dfrac{h}{2}\left[2w(h)+k(t_n)Mh+k(t_{n+1})Mh-k(t_{n+1})M\right]}{\dfrac{h}{2}k(t_n)+K(t_{n+1})} = \frac{0}{k(t_{n+1})} = 0$$

And the second term:

$$\lim_{h \to 0} \frac{\dfrac{h}{2}\left[2w(h)+k(t_n)Mh+k(t_{n+1})Mh-k(t_{n+1})M\right]}{\dfrac{h}{2}k(t_n)+k(t_{n+1})}\left[\left(1+\frac{h}{2}K(t_n)+k(t_{n+1})\right)^n\right]$$

We have:

$$\lim_{h \to 0} \frac{\dfrac{h}{2}\left[2w(h)+k(t_n)Mh+k(t_{n+1})Mh-k(t_{n+1})M\right]}{\dfrac{h}{2}k(t_n)+k(t_{n+1})}\left[\left(1+\frac{h}{2}K(t_n)+k(t_{n+1})\right)^n\right]$$

$$= \lim_{h \to 0} \frac{\dfrac{h}{2}\left[2w(h)+k(t_n)Mh+k(t_{n+1})Mh-k(t_{n+1})M\right]}{\dfrac{h}{2}k(t_n)+k(t_{n+1})} \lim_{h \to 0}\left[\left(1+\frac{h}{2}K(t_n)+k(t_{n+1})\right)^n\right]$$

$$(3.53)$$

The first limit in (3.53) equals zero and:

The computation of $\lim\limits_{h \to 0}\left[\left(1+\dfrac{h}{2}K(t_n)+k(t_{n+1})\right)^n\right]$ is as follows:

Let $y = \lim\limits_{h \to 0}\left[\left(1+\dfrac{h}{2}K(t_n)+k(t_{n+1})\right)^n\right]$ then by tacking the logarithm of the two sides we have:

$$\ln y = \ln \lim_{h \to 0}\left[\left(1+\frac{h}{2}K(t_n)+k(t_{n+1})\right)^n\right] = \lim_{h \to 0}\ln\left[\left(1+\frac{h}{2}K(t_n)+k(t_{n+1})\right)^n\right]$$

$$= \lim_{h \to 0}n\left\{\ln\left[\left(1+\frac{h}{2}K(t_n)+k(t_{n+1})\right)\right]\right\} = \lim_{h \to 0}\frac{t_n-t_0}{h}\left[\ln\left[1+\frac{h}{2}K(t_n)+k(t_{n+1})\right]\right]$$

$$= \lim_{h \to 0}\frac{(t_n-t_0)\left[\ln(1+\frac{h}{2}k(t_n)+k(t_{n+1})\right]}{h}$$

By using the (L'Hospital's Rule):

$$\lim_{h \to 0} \frac{(t_n - t_0) \left[\ln\left(1 + \frac{h}{2} k(t_n) + k(t_{n+1})\right) \right]}{h}$$

$$= \lim_{h \to 0} \frac{(t_n - t_0) \dfrac{1}{\left(1 + \dfrac{h}{2} k(t_n) + k(t_{n+1})\right)} \left[\dfrac{1}{2} k(t_n)\right]}{1}$$

$$= \lim_{h \to 0} \frac{\dfrac{1}{2}(t_n - t_0) k(t_n)}{1 + \dfrac{h}{2} k(t_n) + k(t_{n+1})} = \frac{\dfrac{1}{2}(t - t_0) k(t)}{1 + k(t)}$$

$$(3.54)$$

$\ln y = \dfrac{(t - t_0)k(t)}{2[1 + k(t)]}$ Then $y = e^{\frac{(t-t_0)k(t)}{2[1+k(t)]}}$ hence:

$$\lim_{h \to 0} \left[\left(1 + \frac{h}{2} K(t_n) h + k(t_{n+1})\right)^n \right] = e^{\frac{(t-t_0)k(t)}{2[1+k(t)]}}$$

By substituting in (3.53):

$$\lim_{h \to 0} \frac{\dfrac{h}{2}\left[2w(h) + k(t_n) Mh + k(t_{n+1}) Mh - k(t_{n+1}) M\right]}{\dfrac{h}{2} k(t_n) + k(t_{n+1})}$$

$$\left[\left(1 + \frac{h}{2} K(t_n) + k(t_{n+1})\right)^n\right] = 0 \times e^{\frac{(t-t_0)k(t)}{2[1+k(t)]}} = 0$$

$$(3.55)$$

By substituting from (3.55) and (3.53) in (3.51) then we obtain $\lim_{h \to 0} \|e_{n+1}\| \to 0$ i.e. $\{e_n\}$ converges in m.s to zero as $h \to 0$ hence $X_n \xrightarrow{\text{m.s}} X(t_n) = X(t)$.

SOME RESULTS

Theorem 4.1

Let $\{X_n, n = 0, 1, \cdots\}$, $\{Y_n, n = 0, 1, \cdots\}$ be sequences of 2-r.v's over the same probability space and let a and b be deterministic real numbers.

Suppose: $X = \lim_{n \to \infty} X_n$ and $Y = \lim_{n \to \infty} Y_n$

Then:

1) $(aX + bY) = \lim_{h \to \infty}(aX + bY_n)$

2) $E(X) = \lim_{n \to \infty} E\{X_n\}$

3) $E(XY) = \lim_{n \to \infty} E\{X_n Y_n\}$

4) $\lim_{n \to \infty} E(X_n^2) = E(X^2)$

5) $\lim_{n \to \infty} Var(X_n) == Var(X)$

Definition 4.1: [13]. "The convergence in probability"

A sequence of r.v's $\{X_n\}$ converges in probability to a random variable X as $n \to \infty$ if

$$\lim_{n \to \infty} p\{|X_n - X| > \varepsilon\} = 0 \forall \in 0$$

Definition 4.2 [13]: "The convergence in distribution"

A sequence of r.v's $\{X_n\}$ converge in distribution to a random variable X as $n \to \infty$ if

$$\lim_{n \to \infty} F_{xn}(x) = F_x(x)$$

Lemma (4.1) [13]

The convergence in m.s implies convergence in probability

Lemma (4.2) [13]

The convergence in probability implies convergence in distribution

Theorem 4.2

If $X_n \xrightarrow{m.s} X$ then PDF of $\{X_n\} \xrightarrow{m.s} $ PDF of $\{X\}$ i.e.; $\lim\limits_{n\to\infty} f_{X_n}(x) = f_X(x)$

Proof Since we have shown that If $X_n \xrightarrow{m.s} X$ then

$$X_n \xrightarrow{d} X$$

i.e., if $X_n \xrightarrow{m.s} X$ then $\lim\limits_{n\to\infty} F_{X_n}(x) = F_X(x)$

Then we have: $\lim\limits_{n\to\infty} \dfrac{d}{dx} F_{X_n}(x) = \dfrac{d}{dx} f_X(x)$ then

$$\lim\limits_{n\to\infty} f_{X_n}(x) = f_X(x)$$

NUMERICAL EXAMPLES

Example (5.1)

The differential equation with random term in it and random initial condition

$$y' = Kx, y(x_0) = D, x \in [x_0, x_n],$$

K, D are independent Poisson random variables with joint PDF

$$f_{K,D}(K,D) = \frac{e^{-4} 2^{k+D}}{k!D!}, K,D = 0,1,2,...$$

1. The exact solution,

$$y - D + \frac{K(x^2 - x_0^2)}{2}$$

2. The numerical solution

Using the Random Euler Method:

$$y_1 = y_{n-1} + hf(y_{n-1}, x_{n-1}), y(x_0) = y_0$$

at n = 1

$$y_1 = y_0 + hf(y_0, x_0), D + hKx_0$$

at n = 2

$$y_n = y_1 + hf(y_1, x_1) = D + hKx_0 + hKx_1$$
$$= D + hKx_0 + hK(x_0 + h)$$

at n = 3

$$y_3 = y_2 + hf(y_2, x_2) = D + hKx_0 + hKx_1 + hKx_2$$
$$= D + hKx_0 + hK(x_0 + h) + hK(x_0 + 2h)$$

at n = 4

$$y_4 = y_3 + hf(y_3, x_3) = D + hKx_0 + hKx_1 + hKx_2 + hKx_3$$
$$= D + hKx_0 + hK(x_0 + h) + hK(x_0 + 2h) + hK(x_0 + 3h)$$

and so on…

Then the general numerical solution is

$$y_n = D + hKx_0 + hK(x_0 + h) + hK(x_0 + 2h)$$
$$+ hK(x_0 + 3h) + \cdots + hK(x_0 + (n-1)h)$$

i.e., $y_n = D + hK\sum_{i=0}^{n-1}(x_0 + ih)$.

This can be written in another form:

$$y_n = D + nhKx_0 + [n(n-1)/2]Kh^2.$$

We can prove that:

1. $\lim_{n \to \infty} y_n = y$

Proof Since $\lim\limits_{n\to\infty} y_n = y$ (if and only if) $\lim\limits_{n\to\infty} E|y_n - y|^2 = 0$

Then:

$$y_n - y = nhKx_0 + \frac{n(n-1)}{2}\left[h^2 K\right] - \frac{K\left(x^2 - x_0^2\right)}{2}$$

$$|y_n - y|^2$$

$$= n^2 h^2 K^2 x_0^2 + 2nhK^2 x_0\left[\frac{n(n-1)}{2}h^2 - \frac{\left(x^2 - x_0^2\right)}{2}\right]$$

$$+ K^2\left[\frac{n(n-1)}{2}h^2 - \frac{\left(x^2 - x_0^2\right)}{2}\right]^2$$

Where $h = \dfrac{x_n - x_0}{n}$

$$|y_n - y|^2 = K^2 x_0^2\left(x_n - x_0\right)^2 + K^2 x_0\left(x_n - x_0\right)\left[\frac{(n-1)}{n}\left(x_n - x_0\right)^2 - \left(x^2 - x_0^2\right)\right]$$

$$+ \frac{K^2}{4}\left[\frac{(n-1)}{n}\left(x_n - x_0\right)^2 - \left(x^2 - x_0^2\right)\right]^2$$

$$= K^2 x_0^2 x_n^2 - 2K^2 x_n x_0^3 + K^2 x_0^4 + K^2 x_n x_0\left(\frac{n-1}{n}\right)\left(x_n - x_0\right)^2 - K^2 x_0^2\left(\frac{n-1}{n}\right)\left(x_n - x_0\right)^2$$

$$- K^2 x_n x_0\left(x^2 - x_0^2\right) + K^2 x_0^2\left(x^2 - x_0^2\right) + \frac{K^2}{4}\left(\frac{n-1}{n}\right)^2\left(x_n - x_0\right)^4$$

$$- \frac{K^2}{2}\left(\frac{n-1}{n}\right)\left(x_n - x_0\right)^2\left(x^2 - x_0^2\right) + \frac{K^2}{4}\left(x^2 - x_0^2\right)^2$$

$$\lim_{n\to\infty} E\left|y_n - y\right|^2 = 18x_0^2 x^2 - 36x_0^3 x + 18x_0^4 + 18xx_0 \left(x - x_0\right)^2 - 18x_0^2 \left(x - x_0\right)^2$$

$$- 18xx_0 \left(x^2 - x_0^2\right) + 18x_0^2 \left(x^2 - x_0^2\right) + \frac{18}{4}x^4 - 18x^3 x_0 + 27x^2 x_0^2 - 18xx_0^3$$

$$+ \frac{18}{4}x_0^4 - 9x^4 + 18x^3 x_0 - 18xx_0^3 + 9x_0^4 + \frac{18}{4}x^4 - 9x^2 x_0^2 + \frac{18}{4}x_0^4$$

$$= 18x_0^2 x^2 - 36x_0^3 x + 18x_0^4 + 72xx_0^3 - 36x^2 x_0^2 - 36x_0^4 + \frac{18}{4}x^4 - 18x^3 x_0$$

$$+ 27x^2 x_0^2 - 18xx_0^3 + \frac{18}{4}x_0^4 - 9x^4 + 18x^3 x_0 - 18xx_0^3 + 9x_0^4$$

$$+ \frac{18}{4}x^4 - 9x^2 x_0^2 + \frac{18}{4}x_0^4 = 0$$

i.e; $\lim_{n\to\infty} y_n = y$

We can verify theorem (4.1) as follows:

2. $\lim_{x\to\infty} E\{y_n\} = E\{y\}$

Proof

$$y_n = D + hK\sum_{i=0}^{n-1}(x_0 + ih) = D + hK\sum_{j=1}^{n}(x_0 + (j-1)h) = D + hK\left[x_0 + (x_0 + h) + (x_0 + 2h) + (x_0 + 3h) + \cdots (x_0 + (n-1)h)\right]$$

$$= D + hK\left[nx_0 + h\left[1 + 2 + 3 + \cdots + (n-1)\right]\right] = D + hK\left[nx_0 + h\left(\frac{n(n-1)}{2}\right)\right]$$

$$= D + K\left[nhx_0 + h^2\left(\frac{n(n-1)}{2}\right)\right] \text{ where } h = \frac{x_n - x_0}{n}$$

$$= D + K\left[nx_0 + \frac{(x_n - x_0)}{n} + \left(\frac{(x_n - x_0)}{n}\right)^2\left(\frac{n(n-1)}{2}\right)\right] = D + K\left[x_0(x_n - x_0) + \frac{(n-1)}{2n}(x_n - x_0)^2\right]$$

$$E(y_n) = E(D) + E\left\{K\left[x_0(x_n - x_0) + \frac{(n-1)}{2n}(x_n - x_0)^2\right]\right\} = 2 + 2\left\{E(x_0 x_n) - E(x_0^2) + \frac{(n-1)}{2n}E(x_n - x_0)^2\right\}$$

Then:

$$\lim_{n\to\infty} E\{y_n\} = 2 + 2\left\{E(x_0 x) - E(x_0^2) + \frac{1}{2}E(x - x_0)^2\right\}$$

$$= 2 + 2\left\{E(x_0 x) - E(x_0^2) + \frac{1}{2}E(x^2) - E(x_0 x) + \frac{1}{2}E(x_0^2)\right\}$$

$$= 2 + 2\left\{\frac{1}{2}E(x^2) - \frac{1}{2}E(x_0^2)\right\} = 2 + \left\{E(x^2) - E(x_0^2)\right\} = E\{y\}$$

i.e.; $\lim_{x \to \infty} E\{y_n\} = E\{y\}$

3. $\lim_{n \to \infty} E\{y^2_n\} = E\{y^2\}$

Proof Since $y_n = D + nhKx_0 + [n(n-1)/2]Kh^2$

Then we have:

$$E\{y_n^2\} = E\left[D + Kx_0(x_n - x_0) + Kn(n-1)/2\right]\left[(x_n - x_0)^2 / n^2\right]^2$$

$$= E\left[D + K(x_n - x_0)x_0\right]^2 + 2E\left\{\left[D + K(x_n - x_0)x_0\right]\left[K[(n-1)/n]\left[(x_n - x_0)^2 / 2\right]\right]\right\}$$

$$+ E\left[K[n(n-1)/n]\left[(x_n - x_0)^2 / 2\right]\right]^2$$

$$= E\left[D^2\right] + E\left[K(x_n - x_0)x_0\right]^2 + 2E\left[DK(x_nx_0 - x_0^2)\right] + 2E\left\{\left[D + K(x_n - x_0)x_0\right]\left[K[(n-1)/n]\left[(x_n - x_0)^2 / 2\right]\right]\right\}$$

$$+ E[K^2\left[n(n-1)/n\right]^2 E\left[(x_n - x_0)^2 / 2\right]^2$$

$$= E\left[D^2\right] + E\left[K^2\right]E\left[x_nx_0\right]^2 - 2E\left[K^2\right]E\left[x_nx_0^3\right] + E\left[K^2\right]E\left[x_0\right]^4 + 2E[D]E[K]E\left[x_nx_0\right] - 2E[D]E[K]E\left[x_0^2\right]$$

$$+ 2E[D]E[K]\left[(n-1)/n\right]E\left[(x_n - x_0)^2 / 2\right] + 2\left[(n-1)/n\right]E\left[K^2\right]E\left[x_0(x - x_0)^3 / 2\right]$$

$$+ [n(n-1)/n]^2 E\left[K^2\right]E\left[(x_n - x_0)^2 / 2\right]^2$$

$$= E[D]^2 + E\left[K^2\right]E\left[x_nx_0\right]^2 - 2E\left[K^2\right]E\left[x_nx_0^3\right] + E\left[K^2\right]E\left[x_0\right]^4 + 2E[D]E[K]E\left[x_nx_0\right] - 2E[D]E[K]E\left[x_0^2\right]$$

$$+ E[D]E[K]\left[(n-1)/n\right]E\left[(x_n - x_0)^2\right] + \left[(n-1)/n\right]E\left[K^2\right]E\left[x_0(x - x_0)^3\right] + \frac{1}{4}\left[n(n-1)/n\right]^2 E\left[K^2\right]E\left[(x_n - x_0)^4\right]$$

Then by taking the limit:

$$\lim_{n \to \infty} E\{y_n^2\} = 6 + 6E\left[xx_0\right]^2 - 12E\left[xx_0^3\right] + 6E\left[x_0\right]^4 + 8E\left[xx_0\right] - 8E\left[x_0^2\right]$$

$$+ 4E(x - x_0)^2 + 6E\left[x_0(x - x_0)^3\right] + \frac{6}{4}E(x - x_0)^4$$

$$= 6 + 6E\left[xx_0\right]^2 - 12E\left[xx_0^3\right] + 6E\left[x_0\right]^4 + 8E\left[xx_0\right] - 8E\left[x_0^2\right] + 4E\left[x^2\right]$$

$$- 8E\left[xx_0\right] + 4E\left[x_0^2\right] + 6E\left[x^3x_0\right] - 18E\left[x^2x_0^2\right] + 18E\left[xx_0^3\right] - 6E\left[x_0^4\right]$$

$$+ \frac{6}{4}E\left[x^4\right] - 6E\left[x^3x_0\right] - 6E\left[xx_0^3\right] + 9E\left[x^2x_0^2\right] + \frac{6}{4}E\left[x_0^4\right] = E\left(y^2\right)$$

i.e $\lim_{n \to \infty} E\{y^2_n\} = E\{y^2\}$.

4). $\lim_{n \to \infty} var\{y_n\} = Var\{y\}$

Proof

$$\lim_{n \to \infty} Var\{y_n\} = \lim_{n \to \infty}\left[E\left(y_n^2\right)-\left[E\left(y_n\right)\right]^2\right]$$

$$= \lim_{n \to \infty} E\left(y_n^2\right)-\lim_{n \to \infty}\left[E\left(y_n\right)\right]^2$$

$$= E\left(y^2\right)-\left[E\left(y\right)\right]^2 = Var\left(y\right)$$

i.e., $\lim_{n \to \infty} var\{y_n\} = Var\{y\}$

5). $\lim_{n \to \infty} PDF(y_n) = PDF()$

Proof Since $y = D + \dfrac{K(x^2 - x_0^2)}{2}$

Let us define $Z = D$. Then the inverse transformation is: $y = D + \dfrac{K(x^2 - x_0^2)}{2}$, $D = Z$ then we have $D = Z$ and

$$K = \frac{2(y - Z)}{x^2 - x_0^2}$$

$$J = \begin{vmatrix} \dfrac{\partial k}{\partial y} & \dfrac{\partial k}{\partial z} \\ \dfrac{\partial D}{\partial y} & \dfrac{\partial D}{\partial z} \end{vmatrix} = \begin{vmatrix} \dfrac{2}{x^2 - x_0^2} & \dfrac{-2}{x^2 - x_0^2} \\ 0 & 1 \end{vmatrix} = \frac{2}{x^2 - x_0^2}$$

Then:

$$f_{y,z}\left(y,z\right) = f_{K,D}\left(\frac{2\left(y - Z\right)}{x^2 - x_0^2}, Z\right)|J| = \frac{e^{-4} 2^{Z + \frac{2(Y - Z)}{x^2 - x_0^2}}}{Z!\left[\dfrac{2\left(y - Z\right)}{x^2 - x_0^2}\right]!}$$

Since $D \geq 0$ then $Z \geq 0$ hence

$$y \geq Z + \frac{K(x^2 - x_0^2)}{2}$$

$$f_y(y) = \sum_{Z=0}^{y} \frac{e^{-4} 2^{Z + \frac{2(Y-Z)}{x^2 - x_0^2}}}{Z! \left[\frac{2(y-Z)}{x^2 - x_0^2} \right]!} = PDF(y).$$

For a numerical solution:

Since $y_n = D + nhKx_0 + [n(n-1)/2]Kh^2$

Let $Z_n = D$ then $K = \dfrac{(y_n - z_n)}{nhx_0 + [n(n-1)/2]h^2}$

$$J = \begin{vmatrix} \dfrac{\partial K}{\partial y_n} & \dfrac{\partial K}{\partial Z_n} \\[2mm] \dfrac{\partial D}{\partial y_n} & \dfrac{\partial D}{\partial Z_n} \end{vmatrix}$$

$$= \begin{vmatrix} \dfrac{1}{nhx_0 + \left[n(n-1)/2 \right]h^2} & \dfrac{-1}{nhx_0 + \left[n(n-1)/2 \right]h^2} \\[2mm] 0 & 1 \end{vmatrix}$$

$$= \frac{1}{nhx_0 + \left[n(n-1)/2 \right]h^2}$$

Then:

$$f_{y_n, z_n}(y_n, z_n) = f_{K,D}\left(\frac{(y_n - z_n)}{nhx_0 + \left[n(n-1)/2 \right]h^2}, z_n \right) |J|$$

$$= \frac{e^{-4} 2^{Z_n + \frac{(Y_n - Z_n)}{nhx_0 + [n(n-1)/2]h^2}}}{z_n! \left[\dfrac{(y_n - z_n)}{nhx_0 + \left[n(n-1)/2 \right]h^2} \right]!}$$

$$f_{yn}(y_n) = \sum_{Z_n = 0}^{y_n} \frac{e^{-4} 2^{Z_n + \frac{(Y_n - z_n)}{nhx_0 + [n(n-1)/2]h^2}}}{z_n! \left[\dfrac{(y_n - z_n)}{nhx_0 + \left[n(n-1)/2 \right]h^2} \right]!}$$

Where $h = \dfrac{x_n - x_0}{n}$

$$f_{yn}(y_n) = \sum_{z_n}^{y_n} \frac{e^{-4} 2^{z_n + \frac{(Y_n - z_n)}{(x_n - x_0)x_0 + [(n-1)/2n](x_n - x_0)^2}}}{z_n! \left[\dfrac{(y_n - z_n)}{(x_n - x_0)x_0 + [(n-1)/2n](x_n - x_0)^2} \right]!}$$

$$= PDF(y_n)$$

Then by taking the limit we have

$$\lim_{n \to \infty} f_{yn}(y_n) = \sum_{Z=0}^{y} \frac{e^{-4} 2^{z + \frac{2(Y-z)}{x^2 - x_0^2}}}{z! \left[\dfrac{2(y-z)}{x^2 - x_0^2} \right]!} = f_y(y)$$

i.e.; $\lim\limits_{n \to \infty} PDF(y_n) = PDF(y)$

B. Using the Random Runge-Kutta method:

$$y_{n+1} = y_n + \frac{h}{2}\left[f(y_n, x_n) + f\big(y_n + f(y_n, x_n), x_{n+1}\big) \right],$$

at n = 0

$$y_1 = y_0 + \frac{h}{2}\left[f(y_0, x_0) + f\big(y_0 + f(y_0, x_0), x_1\big) \right]$$

$$= D + \frac{h}{2}\left[Kx_0 + Kx_1 \right] = D + \frac{hK}{2}\left[x_0 + x_1 \right].$$

At n = 1

$$y_2 = y_1 + \frac{h}{2}\left[f(y_1, x_1) + f\big(y_1 + f(y_1, x_1), x_2\big) \right]$$

$$= D + \frac{h}{2}\left[Kx_0 + Kx_1 \right] + \frac{h}{2}\left[Kx_1 + Kx_2 \right]$$

$$= D + \frac{hK}{2}\left[x_0 + 2x_1 + x_2 \right].$$

At n = 2

$$y_3 = y_2 + \frac{h}{2}\left[f(y_2, x_2) + f(y_2 + f(y_2, x_2), x_3) \right]$$

$$= D + \frac{hK}{2}\left[x_0 + 2x_1 + x_2 \right] + \frac{h}{2}\left[Kx_2 + Kx_3 \right]$$

$$= D + \frac{hK}{2}\left[x_0 + 2x_1 + 2x_2 + x_3 \right].$$

Then the general solution is:

$$y_n = D + \frac{hK}{2}\left[x_0 + 2x_1 + 2x_2 + 2x_3 + \ldots\ldots + 2x_{n-1} + x_n \right].$$

$$y_n = D + hK\left[\frac{1}{2}x_0 + (x_0 + h) + (x_0 + 2h) \right.$$
$$\left. + \cdots + (x_0 + (n-1)h) + \frac{1}{2}(x_0 + nh) \right]$$

$$y_n = D + hK\left[\frac{1}{2}x_0 + (x_0 + h) + (x_0 + 2h) \right.$$
$$\left. + \cdots + (x_0 + (n-1)h) + \frac{1}{2}(x_0 + nh) \right]$$

$$y_n = D + hK\sum_{i=0}^{n-1}(x_0 + ih) + \frac{1}{2}nh^2K .$$

This can be written in another form:

$$y_n = D + nhKx_0 + \frac{n^2h^2K}{2}$$

We can prove that:

1. $\lim_{n\to\infty} y_n = y$

Proof:

Since $\lim_{n\to\infty} y_n = y$ (if and only if) $\lim_{n\to\infty} E|y_n - y|^2 = 0$

$$y_n - y = nhKx_0 + \frac{n^2}{2}\left[h^2 K\right] - \frac{K\left(x^2 - x_0^2\right)}{2}$$

$$\left|y_n - y\right|^2$$

$$= n^2 h^2 K^2 x_0^2 + 2nhK^2 x_0\left[\frac{n^2}{2}h^2 - \frac{\left(x^2 - x_0^2\right)}{2}\right]$$

$$+ K^2\left[\frac{n^2}{2}h^2 - \frac{\left(x^2 - x_0^2\right)}{2}\right]^2$$

Where $\quad h = \dfrac{x_n - x_0}{n}$

$$\left|y_n - y\right|^2 = K^2 x_0^2\left(x_n - x_0\right)^2 + K^2 x_0\left(x_n - x_0\right)\left[\left(x_n - x_0\right)^2 - \left(x^2 - x_0^2\right)\right]$$

$$+ \frac{K^2}{4}\left[\left(x_n - x_0\right)^2 - \left(x^2 - x_0^2\right)\right]^2$$

$$= K^2 x_0^2 x_n^2 - 2K^2 x_n x_0^3 + K^2 x_0^4 + K^2 x_n x_0\left(x_n - x_0\right)^2 - K^2 x_0^2\left(x_n - x_0\right)^2$$

$$- K^2 x_n x_0\left(x^2 - x_0^2\right) + K^2 x_0^2\left(x^2 - x_0^2\right) + \frac{K^2}{4}\left(x_n - x_0\right)^4$$

$$- \frac{K^2}{2}\left(x_n - x_0\right)^2\left(x^2 - x_0^2\right) + \frac{K^2}{4}\left(x^2 - x_0^2\right)^2$$

$$\lim_{n\to\infty} E\left|y_n - y\right|^2$$

$$= 18x_0^2 x^2 - 36x_0^3 x + 18x_0^4 + 18xx_0\left(x - x_0\right)^2 - 18x_0^2\left(x - x_0\right)^2 - 18xx_0\left(x^2 - x_0^2\right) + 18x_0^2\left(x^2 - x_0^2\right)$$

$$+ \frac{18}{4}x^4 - 18x^3 x_0 + 27x^2 x_0^2 - 18xx_0^3 + \frac{18}{4}x_0^4 - 9x^4 + 18x^3 x_0 - 18xx_0^3 + 9x_0^4 + \frac{18}{4}x^4 - 9x^2 x_0^2 + \frac{18}{4}x_0^4$$

$$= 18x_0^2 x^2 - 36x_0^3 x + 18x_0^4 + 72xx_0^3 - 36x^2 x_0^2 - 36x_0^4 + \frac{18}{4}x^4 - 18x^3 x_0 + 27x^2 x_0^2 - 18xx_0^3$$

$$+ \frac{18}{4}x_0^4 - 9x^4 + 18x^3 x_0 - 18xx_0^3 + 9x_0^4 + \frac{18}{4}x^4 - 9x^2 x_0^2 + \frac{18}{4}x_0^4$$

i.e.; $\lim\limits_{n\to\infty} y_n = y$

Verification of Theorem (4.1):

2. $\lim_{n\to\infty} E\{y_n\} E\{y\}$

Proof

$$y_n = D + hK\sum_{i=0}^{n-1}(x_0 + ih) + \frac{nh^2 K}{2} = D + hK\sum_{j=1}^{n}(x_0 + (j-1)h) + \frac{nh^2 K}{2}$$

$$= D + hK\left[x_0 + (x_0 + h) + (x_0 + 2h) + (x_0 + 3h) + \cdots(x_0 + (n-1)h)\right] + \frac{nh^2 K}{2}$$

$$= D + hK\left[nx_0 + h\left[1 + 2 + 3 + \cdots + (n-1)\right]\right] + \frac{nh^2 K}{2}$$

$$= D + hK\left[nx_0 + h\left(\frac{n(n-1)}{2}\right)\right] + \frac{nh^2 K}{2}$$

$$= D + K\left[nhx_0 + h^2\left(\frac{n(n-1)}{2}\right)\right] + \frac{nh^2 K}{2} \quad \text{where } h = \frac{x_n - x_0}{n}$$

$$= D + K\left[nx_0\frac{(x_n - x_0)}{n} + \left(\frac{(x_n - x_0)}{n}\right)^2\left(\frac{n(n-1)}{2}\right)\right] + \frac{(x_n - x_0)^2 K}{2n}$$

$$= D + K\left[x_0(x_n - x_0) + \frac{(n-1)}{2n}(x_n - x_0)^2\right] + \frac{(x_n - x_0)^2 K}{2n}$$

$$E(y_n) = E(D) + E\left\{K\left[x_0(x_n - x_0) + \frac{(n-1)}{2n}(x_n - x_0)^2\right] + \frac{(x_n - x_0)^2 K}{2n}\right\}$$

$$= 2 + 2\left\{E(x_0 x_n) - E(x_0^2) + \frac{(n-1)}{2n}E(x_n - x_0)^2\right\} + \frac{2E(x_n - x_0)^2}{2n}$$

$$\lim_{n\to\infty} E\{y_n\} = 2 + 2\left\{E(x_0 x) - E(x_0^2) + E(x - x_0)^2 / 2\right\}$$

$$= 2 + 2\left\{E(x_0 x) - E(x_0^2) + \frac{1}{2}E(x^2) - E(x_0 x) + \frac{1}{2}E(x_0^2)\right\}$$

$$= 2 + E(x^2) - E(x_0^2) = E\{y\}$$

i.e.; $\lim_{n \to \infty} E\{y_n\} E\{y\}$.

3. $\lim_{n \to \infty} E\{y_n^2\} E\{y^2\}$

Proof since $y_n = D + nhKx_0 + \dfrac{n^2 h^2 K}{2}$ then:

$$E\{y_n^2\} = E\left[D + Kx_0(x_n - x_0) + \frac{(x_n - x_0)^2 K}{2}\right]^2$$

$$= E\left[D + K(x_n - x_0)x_0\right]^2 + 2E\left\{\left[D + K(x_n - x_0)x_0\right]\left[\frac{K(x_n - x_0)^2}{2}\right]\right\} + E\left[\frac{K(x_n - x_0)^2}{2}\right]^2$$

$$= E\left[D^2\right] + E\left[K(x_n - x_0)x_0\right]^2 + 2E\left[DK(x_n x_0 - x_0^2)\right]$$

$$+ 2E\left\{\left[D + K(x_n - x_0)x_0\right]\left[\frac{K(x_n - x_0)^2}{2}\right]\right\} + E\left[\frac{K(x_n - x_0)^2}{2}\right]^2$$

$$= E\left[D^2\right] + E\left[K^2\right]E\left[x_n x_0\right]^2 - 2E\left[K^2\right]E\left[x_n x_0^3\right] + E\left[K^2\right]E\left[x_0\right]^4$$

$$+ 2E[D]E[K]E\left[x_n x_0\right] - 2E[D]E[K]E\left[x_0^2\right] + 2E[D]E[K]E\left[(x_n - x_0)^2/2\right]$$

$$+ 2E\left[K^2\right]E\left[x_0(x - x_0)^3/2\right] + E\left[K^2\right]E\left[(x_n - x_0)^2/2\right]^2$$

$$= E[D]^2 + E\left[K^2\right]E\left[x_n x_0\right]^2 - 2E\left[K^2\right]E\left[x_n x_0^3\right] + E\left[K^2\right]E\left[x_0\right]^4$$

$$+ 2E[D]E[K]E\left[x_n x_0\right] - 2E[D]E[K]E\left[x_0^2\right] + E[D]E[K]E\left[(x_n - x_0)^2\right]$$

$$+ E\left[K^2\right]E\left[x_0(x_n - x_0)^3\right] + \frac{1}{4}E\left[K^2\right]E\left[(x_n - x_0)^4\right]$$

$$\lim_{n \to \infty} E\{y_n^2\} = 6 + 6E\left[xx_0\right]^2 - 12E\left[xx_0^3\right] + 6E\left[x_0\right]^4 + 8E\left[xx_0\right] - 8E\left[x_0^2\right]$$

$$+ 4E\left[(x - x_0)^2\right] + 6E\left[x_0(x - x_0)^3\right] + \frac{6}{4}E\left[(x - x_0)^4\right]$$

$$= 6 + 6E\left[xx_0\right]^2 - 12E\left[xx_0^3\right] + 6E\left[x_0\right]^4 + 8E\left[xx_0\right] - 8E\left[x_0^2\right] + 4E\left[x^2\right]$$

$$- 8E\left[xx_0\right] + 4E\left[x_0^2\right] + 6E\left[x^3 x_0\right] - 18E\left[x^2 x_0^2\right] + 18E\left[xx_0^3\right] - 6E\left[x_0^4\right]$$

$$+ \frac{6}{4}E\left[x^4\right] - 6E\left[x^3 x_0\right] - 6E\left[xx_0^3\right] + 9E\left[x^2 x_0^2\right] + \frac{6}{4}E\left[x_0^4\right] = E\left(y^2\right)$$

i.e.; $\lim_{n\to\infty} E\{y_n^2\} E\{y^2\}$

4. $\lim_{n\to\infty} Var(y_n) = Var(y)$

Proof

$$\lim_{n\to\infty} Var\{y_n\} = \lim_{n\to\infty}\left[E\left(y_n^2\right)-\left[E\left(y_n\right)\right]^2\right]$$

$$= \lim_{n\to\infty} E\left(y_n^2\right)-\lim_{n\to\infty}\left[E\left(y_n\right)\right]^2$$

$$= E\left(y^2\right)-\left[E\left(y\right)\right]^2 = Var\left(y\right)$$

i.e. $\lim_{n\to\infty} Var(y_n) = Var(y)$

5. $\lim_{n\to\infty} PDF(y_n) = PDF(y)$

Since $y = D + \dfrac{K(x^2 - x_0^2)}{2}$

Let us define $Z = D$. Then the inverse transformation is:

$D = y + \dfrac{K(x^2 - x_0^2)}{2}$ $D = z$ then we have $D = z$ and $K = \dfrac{2(y - Z)}{x^2 - x_0^2}$

$$J = \begin{vmatrix} \dfrac{\partial K}{\partial y} & \dfrac{\partial K}{\partial z} \\ \dfrac{\partial D}{\partial y} & \dfrac{\partial D}{\partial z} \end{vmatrix} = \begin{vmatrix} \dfrac{2}{x^2 - x_0^2} & \dfrac{-2}{x^2 - x_0^2} \\ 0 & 1 \end{vmatrix} = \dfrac{2}{x^2 - x_0^2}$$

$$f_{y,z}\left(y,z\right) = f_{K,D}\left(\dfrac{2(y-z)}{x^2 - x_0^2}, z\right)|J| = \dfrac{e^{-4}2^{Z+\frac{2(Y-Z)}{X^2-X_0^2}}}{z!\left[\dfrac{2(y-z)}{x^2 - x_0^2}\right]!}$$

Since $D \geq 0$ then $z \geq 0$ this implies

$y \geq z + \dfrac{K(x^2 - x_0^2)}{2}$

$$f_y(y) = \sum_{z=0}^{y} \frac{e^{-4}2^{Z+\frac{2(Y-Z)}{X^2-X_0^2}}}{z!\left[\frac{2(y-z)}{x^2-x_0^2}\right]!} = PDF(y).$$

.

Numerically since $y_n = D + nhKx_0 + \dfrac{n^2h^2K}{2}$

Let $Z_n = D$ then $K = \dfrac{(y_n - z_n)}{nhx_0 + \dfrac{n^2h^2}{2}}$

$$J = \begin{vmatrix} \dfrac{\partial k}{\partial y_n} & \dfrac{\partial k}{\partial z_n} \\[3mm] \dfrac{\partial D}{\partial y_n} & \dfrac{\partial D}{\partial z_n} \end{vmatrix} = \begin{vmatrix} \dfrac{1}{nhx_0 + \dfrac{n^2h^2}{2}} & \dfrac{-1}{nhx_0 + \dfrac{n^2h^2}{2}} \\[3mm] 0 & 1 \end{vmatrix}$$

$$= \dfrac{1}{nhx_0 + \dfrac{n^2h^2}{2}}$$

$$f_{y_n \cdot z_n}(y_n, z_n) = f_{k,D}\left(\frac{(y_n - z_n)}{nhx_0 + \dfrac{n^2h^2}{2}}, z_n\right)|J|$$

$$= \frac{e^{-4}2^{Z_n + \frac{(Y_n - Z_n)}{nhx_0 + \frac{n^2h^2}{2}}}}{z_n!\left[\dfrac{(y_n - z_n)}{nhx_0 + \dfrac{n^2h^2}{2}}\right]!}$$

$$f_{yn}(y_n) = \sum_{z_n=0}^{y_n} \frac{e^{-4}2^{Z_n + \frac{(Y_n - Z_n)}{nhx_0 + \frac{n^2h^2}{2}}}}{z_n!\left[\dfrac{(y_n - z_n)}{nhx_0 + \dfrac{n^2h^2}{2}}\right]!}$$

Where $h = \dfrac{X_n - X_0}{n}$

$$f_{yn}(y_n) = \sum_{z_n}^{y_n} \frac{e^{-4} 2^{\, z_n + \frac{(Y_n - Z_n)}{(x_n - x_0)x_0 + (x_n - x_0)^2/2}}}{z_n! \left[\dfrac{(y_n - z_n)}{(x_n - x_0)x_0 + (x_n - x_0)^2/2} \right]!}$$

$$= PDF(y_n)$$

$$\lim_{n \to \infty} f_{yn}(y_n) = \sum_{z=0}^{y} \frac{e^{-4} 2^{\, z + \frac{2(Y-Z)}{x^2 - x_0^2}}}{z! \left[\dfrac{2(y-z)}{x^2 - x_0^2} \right]!} = f_y(y)$$

i.e.; $\lim\limits_{n \to \infty} PDF(y_n) = PDF(y)$.

Example (5.2)

Solve the problem

$$\frac{dy}{dt} = y^2 - Ky, \; y(0) = K, \; t \in [0, t_n], \; K \sim \exp(1)$$

The exact solution

Y=K

The numerical solution by the Euler method:

$$y_n = y_{0-1} + hf(y_{n-1}, t_{n-1}) y(0) = K.$$

AT n = 1

$$y_1 = y_0 + hf(y_0, t_0) = K + h(y_0^2 - ky_0) = K.$$

At n = 2

$$y_2 = y_1 + hf(y_1, t_1) = K + h(y_1^2 - ky_1) = K.$$

At n = 3

$$y_3 = y_2 + hf(y_2, t_2) = K + h(y_2^2 - ky_2) = K.$$

And so on....

Then the general numerical solution: $y_n = K$.

It is clear that:

1. $\lim_{n \to \infty} y_n = y$

Since $\lim_{n \to \infty} E|y_n - y|^2 = \lim_{n \to \infty} E|K - K|^2 = 0$

Verification of Theorem (4.1)

It is clear that:

2. $\lim_{n \to \infty} E\{y_n\} E\{y\}$

$E(y_n) = E(K) = K = E(y)$

3. $\lim_{n \to \infty} E\{y_n^2\} E\{y^2\}$

Since $y_n = K \Rightarrow E\{y_n^2\} = E[K]^2 = K^2 E\{y^2\}$

4. $\lim_{n \to \infty} Var(y_n) = Var(y)$

$$\lim_{n \to \infty} Var\{y_n\} = \lim_{n \to \infty} \left[E(y_n^2) - [E(y_n)]^2 \right]$$

$$= \lim_{n \to \infty} E(y_n^2) - \lim_{n \to \infty} [E(y_n)]^2$$

$$= E(y^2) - [E(y)]^2 = Var(y)$$

5. $\lim_{n \to \infty} PDF(y_n) = PDF(y)$

$y = K$ Then $|J| = 1$ which implies

$$PDF(y) = |J| PDF(k) = e^{-y}$$

$y_n = K$ Then $|J| = 1$ which implies

$$PDF(y_n) = |J| PDF(K) = e^{-y_n k}$$

Then: $\lim_{n \to \infty} PDF(y_n) = PDF(y)$

CONCLUSIONS

The initially valued first order random differential equations can be solved numerically using the random Euler and random Runge-Kutta methods in mean square sense. The existence and uniqueness of the solution have been proved. The convergence of the presented numerical techniques has been proven in mean square sense. The results of the paper have been illustrated through some examples.

REFERENCES

1. K. Burrage and P. M. Burrage, "High Strong Order Explicit Runge-Kutta Methods for Stochastic Ordinary Differential Equations," Applied Numerical Mathematics, Vol. 22, No. 1-3, 1996, pp. 81-101. doi:10.1016/S0168-9274(96)00027-X
2. K. Burrage and P. M. Burrage, "General Order Conditions for Stochastic Runge-Kutta Methods for Both Commuting and Non-commuting Stochastic Ordinary Equations," Applied Numerical Mathematics, Vol. 28, No. 2-4, 1998, pp. 161-177. doi:10.1016/S0168-9274(98)00042-7
3. J. C. Cortes, L. Jodar and L. Villafuerte, "Numerical Soluion of Random Differential Equations, a Mean Square Approach," Mathematical and Computer Modelling, Vol. 45, No. 7, 2007, pp. 757-765. doi:10.1016/j.mcm.2006.07.017
4. J. C. Cortes, L. Jodar and L.Villafuerte, "A Random Euler Method for Solving Differential Equations with Uncertainties," Progress in Industrial Mathematics at ECMI, Madrid, 2006.
5. 5.H. Lamba, J. C. Mattingly and A. Stuart, "An adaptive Euler-Maruyama Scheme for SDEs, Convergence and Stability," IMA Journal of Numerical Analysis, Vol. 27, No. 3, 2007, pp. 479-506. doi:10.1093/imanum/drl032
6. E. Platen, "An Introduction to Numerical Methods for Stochastic Differential Equations," Acta Numerica, Vol. 8, 1999, pp. 197-246. doi:10.1017/S0962492900002920
7. D. J. Higham, "An Algorithmic Introduction to Numerical Simulation of SDE," SIAM Review, Vol. 43, No. 3, 2001, pp. 525-546. doi:10.1137/S0036144500378302

8. D. Talay and L. Tubaro, "Expansion of the Global Error for Numerical Schemes Solving Stochastic Differential Equation," Stochastic Analysis and Applications, Vol. 38, No. 4, 1990, pp. 483-509. doi:10.1080/07362999008809220
9. P. M. Burrage, "Numerical Methods for SDE," Ph.D. Thesis, The University of Queensland, 1999.
10. P. E. Kloeden, E. Platen and H. Schurz, "Numerical Solution of SDE Through Computer Experiments," Second Edition, Springer, Berlin, 1997.
11. M. A. El-Tawil, "The Approximate Solutions of Some Stochastic Differential Equations Using Transformation," Journal of Applied Mathematics and Computing, Vol. 164, No. 1, 2005, pp. 167-178. doi:10.1016/j.amc.2004.04.062
12. P. E. Kloeden and E. Platen, "Numeical Solution of Stochastic Differential Equations," Springer, Berlin, 1999.
13. T. T. Soong, "Random Differential Equations in Science and Engineering," Academic Press, New York, 1973.

CITATION

1. M. El-Tawil and M. Sohaly, "Mean Square Numerical Methods for Initial Value Random Differential Equations," Open Journal of Discrete Mathematics, Vol. 1 No. 2, 2011, pp. 66-84. doi: 10.4236/ojdm.2011.12009.

An Introduction to Paraconsistent Integral Differential Calculus: With Application Examples

João Inácio Da Silva Filho[1, 2]
[1]Group of Applied Paraconsistent Logic, Santa Cecília University, Santos, Brazil
[2]Institute for Advanced Studies, University of São Paulo, São Paulo, Brazil

ABSTRACT

In this paper we show that it is possible to integrate functions with concepts and fundamentals of Paraconsistent Logic (PL). The PL is a non-classical Logic that tolerates the contradiction without trivializing its results. In several works the PL in his annotated form, called Paraconsistent logic annotated with annotation of two values (PAL2v), has presented good results in analysis of information signals. Geometric interpretations based on PAL2v-Lattice associate were obtained forms of Differential Calculus to a Paraconsistent Derivative of first and second-order functions. Now, in this paper we extend the calculations for a form of Paraconsistent Integral Calculus that can be viewed through the analysis in the PAL2v-Lattice. Despite improvements that can develop calculations in complex functions, it is verified that the use of Paraconsistent Mathematics in differential and Integral Calculus opens a promising path in researches developed for solving linear and non-linear systems. Therefore the Paraconsistent Integral Differential Calculus can be an important tool in systems by modeling and solving problems related to Physical Sciences.

INTRODUCTION

The Paraconsistent Logic (PL) belongs to the class of non-classical logics and presents in its foundation some tolerances at contradiction, without invalidating the conclusions. Its extended form called Paraconsistent Annotated Logic (PAL), has in its representation an associated Lattice that allows the development of algorithmic techniques and direct applications [1] -[3] . Paraconsistent Mathematics is structured on Paraconsistent Logic (PL) and has as main purpose the study of common mathematical objects such as sets, numbers and functions, where some contradictions are allowed. The PAL treating information signals in its special form called Paraconsistent Logic with annotation of two values (PAL2v) allow extracting equations for applications in signal analysis. It is possible through a Lattice FOUR of values (Hasse Diagram), obtained in the PAL2v representation, how much the annotation (or evidences) can express the knowledge about a proposition P.

The PAL2v-Lattice [4] can be formed ordered pairs of values (m, λ), which will form the annotation. In this representation, an operator ~ is fixed: $|\tau| \rightarrow |\tau|$ where: $\tau = \{(\mu, \lambda) | \mu, \lambda \in [0,1]\} \subset \Re$.

As seen in [3] and [4] through geometric transformations we can find a Lattice of values equivalent to an associate at PAL2v. These interpretations with PAL2-Lattice of values allow Paraconsistent mathematical calculations through equations of parameterization [4]. In x-axis, PAL2v-Lattice is possibly identified by the Certainty Degree that is achieved by:

$$D_C = \mu - \lambda$$

(1)

where: m is Favorable evidence Degree, where $\{\mu \in [0,1]\} \subset \Re$.

λ is Favorable evidence Degree, where $\{\lambda \in [0,1]\} \subset \Re$.

As seen in Figure 1 the Certainty Degree values, which belong to the set Â vary in closed range −1 to +1 and are in the horizontal axis of the PAL2v-Lattice of values called "Axis of degrees of certainty".

The Contradiction Degree (D_{ct}) is obtained by:

$$D_{ct} = \mu + \lambda - 1 \tag{2}$$

As seen in Figure 1 the resulting values of D_{ct} belong to set Â, vary on the closed interval +1 and −1 and are exposed on the vertical axis of the PAL2v-Lattice called "Axis of contradiction degrees".

By analyzing the PAL2v-Lattice [4] -[7] the concept of Paraconsistent logical state (e_τ) can be correlated to the fundamental concept of state, as studied in physical science and then extended to the model based on Paraconsistent Logic.

$$\varepsilon_{\tau(\mu,\lambda)} = (D_C, D_{ct}) \tag{3}$$

where: e_τ is the Paraconsistent logical state.

D_C is the Certainty Degree obtained according to the two degrees of Evidence μ and λ.

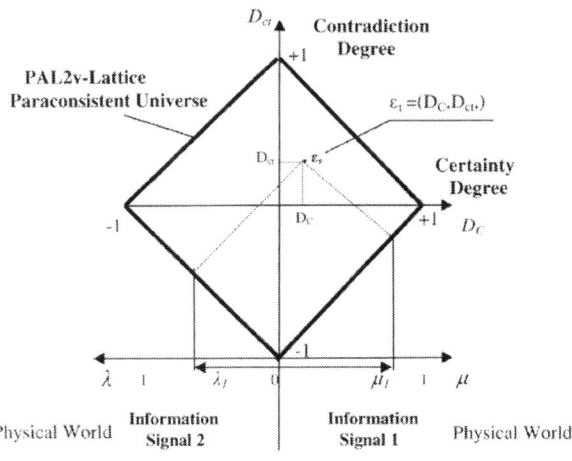

Figure 1: PAL2v-Lattice with Certainty Degree, Contradictory Degree and Paraconsistent logical signal.

D_{ct} is the Contradiction Degree found according to the two degrees of Evidence μ and λ.

Figure 1 shows the Paraconsistent logical state e_τ.

The Certainty Degree normalized [6] [8] from the Paraconsistent Logical Model is called Resulting Degree of Evidence which is calculated by:

$$\mu_{\text{Re}} = \frac{D_C + 1}{2}$$

(4)

Likewise, the normalized Contradiction Degree from the Paraconsistent Logical Model is calculated by:

$$\mu_{ctr} = \frac{D_{Ct} + 1}{2}$$

(5)

Derivative and Newton's Quotient

According to definition, the Derivative [9] -[12] of a function of one variable is defined as a limit process:

$f'(a) = \lim\limits_{x \to a} \dfrac{f(x) - f(a)}{x - a}$. Considering there was an increase h such that:

$h = x - a \leftrightarrow x = a + h$, the Derivative can be rewritten as:

$$f'(a) = \lim\limits_{h \to 0} \frac{f(a+h) - f(a)}{h}$$

(6)

The equation can be written as h represents a variation of x, such that:

$$\lim\limits_{\Delta x \to 0} \frac{\Delta y}{\Delta x} = \lim\limits_{\Delta x \to 0} \frac{f(x + \Delta x) - f(x)}{\Delta x}$$

(7)

Therefore, the Newton's quotient is defined as the incremental ratio of f with respect to the variable x, at the point x.

$$Q(y, \Delta x) = \frac{f(x + \Delta x) - f(x)}{\Delta x}$$

(8)

The PAL treating information signals in its special form called Paraconsistent Logic with annotation of two values (PAL2v) allows extracting equations for applications in signal analysis from the Newton's quotient.

PARACONSISTENT MATHEMATICS

With PAL2v applied in the Newton's quotient, we can obtain all the information necessary and sufficient to effect the derivation of first and second order and apply them to physical systems with good results without ignoring the action of the infinitesimal [13] [14] .

First-Order Paraconsistent Derivative

Initially we will apply to the Newton's quotient a factor of normalization K. This is necessary for that we can put its values within the limits of the PAL2v-Lattice, therefore:

$$Q(y, \Delta x) = \frac{1}{\Delta x}\left(\frac{f(x+\Delta x)}{K} - \frac{f(x)}{K} \right)$$

(9)

where: K is a normalization factor, whose action allows the equation to be done as the fundamentals of PAL2v.

With the normalization factor in Equation (9) are identified Degrees of Evidence of PAL2v annotation, such that:

$$\mu = \frac{f(x+\Delta x)}{K} \rightarrow \text{Favorable Evidence Degree and } \lambda = \frac{f(x)}{K} \rightarrow \text{Unfavor-}$$

able Evidence Degree.

From Equation (2) we have the Certainty Degree of the Newton's quotient:

$$D_{C(N)} = \left[\frac{f(x+\Delta x)}{K} - \frac{f(x)}{K} \right]$$

(10)

Similarly, from the Equation (3) the Contradiction Degree of the Newton's quotient:

$$D_{ct(\psi N)} = \left[\frac{f(x+\Delta x)}{K} + \frac{(x)}{K} - 1 \right]$$

(11)

In Paraconsistent Logical Model the K value must be estimated so that the values of the degrees of evidence become established within the fundamentals of PAL2v. We can adjust this value is an equilibrium point equivalent to Planck's constant called Paraquantum Factor of quantization, as seen in [5] and [7] .

$$K_N = \sqrt{2} y_{max}$$

(12)

Be y_{max} the maximum value of the function at the considered point.

K_N Paraconsistent Newton Normalization Factor.

The value of the Paraconsistent Derivative of the first-order in the physical world is obtained through reapplying the Newton Normalization Factor (K_N) in the result obtained by Paraconsistent Newton's quotient:

$$y' = K_N \times PQ_{(N\psi)}$$

(13)

Thus, it is possible for the Paraconsistent Mathematics to be connected to the equilibrium point, defined by the Paraquantum Factor of quantization (h_ψ) of the PAL2v-Lattice [6] [8] . Therefore, the Paraconsistent values extracted from Newton's quotient adjusted to Paraconsistent Logical Model depends of Δx, that is, the increment of the variable x applied to the calculations. It is verified that the location of the Paraconsistent logical state was adjusted in the PAL2v-Lattice through the Newton Normalization Factor (K_N) and identifies how it is represented any differentiable function f(x) before the mathematical procedure for obtaining the derivative. Thus, this normalization allows the function $y=x^n$ to be identified in the Paraconsistent Newton's quotient, such that:

$$PQ_{(\psi N)} = \left[\frac{(x+\Delta x)^n}{K_N \Delta x} - \frac{(x)^n}{K_N \Delta x} \right]$$

(14)

where: $\mu_{\psi N} = \dfrac{(x+\Delta x)^n}{K_N}$ is the Favorable Evidence Degree and $\lambda_{\psi N} = \dfrac{(x)^n}{K_N}$ is the Unfavorable Evidence Degree. Thus, for the function $y=x^n$, the Paraconsistent Newton's quotient produces the value corresponding to the Certainty Degree (D_c), which, as the fundamentals of PAL2v is obtained by Equation (12):

$$D_{C(\psi N)} = \left[\frac{(x+\Delta x)^n}{K_N} - \frac{(x)^n}{K_N} \right]$$

(15)

Similarly, the Equation (13) the Contradiction Degree of the Paraconsistent Newton's quotient:

$$D_{ct(\psi N)} = \left[\frac{(x+\Delta x)^n}{K_N} + \frac{(x)^n}{K_N} - 1 \right]$$

16)

And from the Equation (4) the Evidence Degree resulting of Paraconsistent Newton's quotient:

$$\mu_{ReQ(\psi N)} = \frac{(x+\Delta x)^n}{2K_N} - \frac{(x)^n}{2K_N} + \frac{1}{2}$$

(17)

And from the Equation (5) the normalized Contradiction Degree of Paraconsistent Newton's quotient:

$$\mu_{ctrQ(\psi N)} = \frac{(x+\Delta x)^n}{2K_N} + \frac{(x)^n}{2K_N}$$

(18)

Example of First-Order Paraconsistent Derivative Application
Calculate the final value of the first-order Paraconsistent Derivative of the function f(x) = x³ in x=2, with $\Delta x = 0.001$.

Resolution: Initially to form Newton Normalization Factor it is calculated the maximum value of the function f(x) = x³ at the point considered x=2, therefore, being:

$$f(x) = x^3 \rightarrow f(x) = y_{max} \rightarrow f(x) = 2^3 = 8$$

The value of the Newton Normalization Factor, according to the Equation (12), is:

$$K_N = \sqrt{2} y_{max} \rightarrow K_N = 8\sqrt{2}$$

The Certainty Degree of Newton's quotient is calculated by Equation (16):

$$D_{CQ(\psi N)} = \frac{(2+0.001)^3}{8\sqrt{2}} - \frac{(2)^3}{8\sqrt{2}}$$

The Paraconsistent Newton's quotient is calculated according to Equation (15):

$$PQ_{(\psi N)} = \frac{1}{0.001}\left[\frac{(2+0.001)^3}{8\sqrt{2}} - \frac{(2)^3}{8\sqrt{2}}\right] \to PQ_{(\psi N)} = \frac{1}{0.001}[0.708167971 - 0.707106781]$$

$$PQ_{(\psi N)} = 1.061190776$$

Recovering the value of Paraconsistent Derivative in the physical world by Equation (13):

$$y' = K_N \times PQ_{(\psi N)} \to y' = 8\sqrt{2} \times 1.061190776 = 12.0060031$$

Then, for these conditions of $\Delta x = 0.001$ the value of the first-order Paraconsistent Derivative of function $f(x) = x^3$ in x=2 is: $y' = 12.0060031$

Paraconsistent Second-Order Derivative
Whereas the first-order Paraconsistent derivative is obtained with the calculation of the Paraconsistent Newton's quotient Equation (9), then the Certainty Degree is:

$$D_{CQ(\psi N)} = \left[\frac{f(x+\Delta x)}{K_N} - \frac{f(x)}{K_N}\right]$$

$$(19)$$

This first value of the Certainty Degree will be normalized by application Equation (4), turning into Favorable Evidence Degree to the second-order Derivative, so:

$$\mu_{2Q(\psi N)} = \frac{D_{C1Q(\psi N)} + 1}{2}$$

$$(20)$$

Or then, (19) in (20), we have:

$$\mu_{2Q(\psi N)} = \frac{\left[\dfrac{f(x+\Delta x)}{K_N} - \dfrac{f(x)}{K_N}\right] + 1}{2}$$

$$(21)$$

The equation of Paraconsistent Newton's quotient of the second point, or second Paraconsistent logical state, obtained into PAL2v-Lattice for second-order derivative is:

$$PQ_{(\psi N)2} = \frac{1}{\Delta x}\left(\frac{f(x)}{K_N} - \frac{f(x-\Delta x)}{K_N} \right)$$

(22)

Are identified in the Equation (22) the degrees of evidence, such that:

$$\mu_{2\psi} = \frac{f(x)}{K_N} \rightarrow \text{Second Favorable Evidence Degree, and } \lambda_{2\psi} = \frac{f(x-\Delta x)}{K_N} \rightarrow$$

Second Unfavorable Evidence Degree. The Certainty Degree of second Paraconsistent logical state is calculated by:

$$D_{C2Q(\psi N)} = \frac{f(x)}{K_N} - \frac{f(x-\Delta x)}{K_N}$$

(23)

The second value of the Certainty Degree will be normalized, thus becoming by Equation (4) in Unfavorable Evidence Degree to the second-order Derivative of the same function f(x), so:

$$\lambda_{2Q(\psi N)} = \frac{D_{C2Q(N)}+1}{2}$$

(24)

Or then, (23) in (24), we have:

$$\lambda_{2Q(\psi N)} = \frac{\left[\frac{f(x)}{K_N} - \frac{f(x-\Delta x)}{K_N} \right]+1}{2}$$

(25)

For this second representation of Paraconsistent Derivative when decreases the value of Δx the Unfavorable Evidence degree λ_2 approach of the Favorable Evidence Degree μ_2. Thus, the Paraconsistent Derivative of second order will be:

$$PQ^2_{(\psi N)} = \frac{1}{\Delta x}\left[\frac{\mu_{2Q(\psi N)} - \lambda_{2Q(\psi N)}}{\Delta x} \right]$$

(26)

The analysis of sequence in PAL2v will result in the Certainty degree divided by the value of the square of the increase of the variable x, so:

$$PQ^2_{(\psi N)} = \frac{\mu_{2Q(\psi N)} - \lambda_{2Q(\psi N)}}{(\Delta x)^2} = \frac{D_{C3Q(\psi N)}}{(\Delta x)^2}$$

The Equations (20) and (24) in (26), results in:

$$PQ^2_{(\psi N)} = \frac{1}{\Delta x} \times \frac{\dfrac{D_{C1Q(\psi N)}+1}{2} - \dfrac{D_{C2Q(\psi N)}+1}{2}}{\Delta x}.$$

Or, making (21) and (25) in (26) and rearranging, the Paraconsistent Newton's quotient for second-order function is:

$$PQ^2_{(\psi N)} = \frac{1}{2(\Delta x)^2}\left[\left(\frac{f(x+\Delta x)}{K_N} - \frac{f(x)}{K_N}\right) - \left(\frac{f(x)}{K_N} - \frac{f(x-\Delta x)}{K_N}\right)\right]$$

(27)

where: $PQ^2_{(\psi N)}$ final value of the Paraconsistent Derivative function second-order.

K_N is the Normalization Newton factor.

To recover and so obtain the Paraconsistent Derivative value for second-order function f(x) in actual physical universe:

$$y'' = 2 \times K_N \times PQ^2_{(\psi N)}$$

(28)

where: y″ is the second-order Derivative in real world.

Example of Second-Order Paraconsistent Derivative Application
Calculate the final value of the second-order Paraconsistent Derivative of the function:

$f(x) = x^3$ in x=3, with $\Delta x = 0.001$.

For the resolution, initially is estimated the maximum function value at the point considered x=3.

Therefore, being:

$$f(x) = x^3 \rightarrow f(x) = y_{\max} \rightarrow f(x) = 3^3 = 27.$$

The Paraconsistent Newton Normalization Factor is calculated by Equation (12):

$$K_N = \sqrt{2}.y_{\max} \rightarrow K_N = 27\sqrt{2}$$

With the Equation (27) is obtained the second-order Paraconsistent Derivative, that with: $\Delta x = 0.001$ it comes:

$$PQ^2_{(\psi N)} = \frac{1}{2 \times 27 \times \sqrt{2}\,(0.001)^2}\left[\left(f(3+0.001)^3 - f(3)^3\right) - \left(f(3)^3 - f(3-0.001)^3\right)\right]$$

$$PQ^2_{(\psi N)} = \frac{1}{2 \times 27 \times \sqrt{2}\,(0.001)^2}\left[(27.027009 - 27) - (27 - 26.973009)\right]$$

$$PQ^2_{(\psi N)} = \frac{1}{2 \times 27 \times \sqrt{2}\,(0.001)^2}\left[(0.027009) - (0.026991001)\right]$$

$$PQ^2_{(\psi N)} = \frac{0.000017999}{0.000076367} \rightarrow PQ^2_{(\psi N)} = 0.235689165$$

The value of the second-order Paraconsistent Derivative of function f(x) in actual physical universe is obtained by applying the Equation (28):

$$y'' = 2 \times K_N \times PQ^2_{(\psi N)} = 2 \times 27\sqrt{2} \times 0.235689165 = 17.999$$

Then, for these conditions of $\Delta x = 0.001$ the value of the second-order Paraconsistent Derivative of function $f(x) = x^3$ in x=3 is: $y'' = 17.999$.

PARACONSISTENT INTEGRAL CALCULUS

In the calculations of Derivative every one of the Primitive functions, called here by F(x), corresponds to a Derivative function $F' = (x)$ [11] [14]. Is seen also that any Primitive function plus a constant positive or negative is another same Primitive Derivative, so we can establish a General Primitive function type: $F(x) = F_0(x) + C$

The General Primitive function F(x) is called the Indefinite Integral of the Differential dF(x). Therefore, the Indefinite Integral of f(x)dx is represented symbolically by: $\int f(x)dx = F(x)$

$$\int f(x)dx = F_0(x) + C$$

(29)

By other side, the Definite Integral is as an insertion of a function and extraction a number, whose value corresponds to the area between the graph of the function and the axis of x. In the calculation of Definite Integral are established the limits of integration, so the calculation is a mathematical process established between two well-defined intervals [14] [15].

The application of the concept of Integration in a function through Paraconsistent Logical Model will be made based on the Derivative process that uses the incremental rate, or Newton's quotient. For this condition the Paraconsistent logical State from Equation (3), which is defined in the PAL2v-Lattice by the values of the degrees of evidence, is located at one point represented by the Certainty Degree (Equation (15)) and the Contradiction Degree (Equation (16)) of Paraconsistent Newton's quotient:

$$\varepsilon_\tau = \left(D_{C(\psi N)}, D_{ct(\psi N)} \right)$$

(30)

This means that for any type function y = f(x) where we can obtain the incremental ratio or Newton's quotient, its Derivative with the adjustment with Newton normalization factor (K_N) is obtained with the representation of Paraconsistent logical state onto axis of the degrees of contradiction. Figure 2 shows the location in the PAL2v-Lattice of Paraconsistent logical state $\varepsilon_{h\psi}$.

In the method of integration in conventional mode leads to ignore the infinitesimal, as also is made in the method of limits, when considered the increase of variable x tends to zero.

In the conventional integral method for the function $y = x^n$ it is seen that: To a function of type $y = x^n$ where n is any positive integer, the initial analysis is done via the binomial theorem [14] [15], where the term $(x + \Delta x)^n$ located on the left of the numerator of the Newton's quotient in Equation (14), it is written as:

$$(x + \Delta x)^n = x^n + nx^{n-1}(\Delta x) + \frac{n(n-1)}{2!}x^{n-2}(\Delta x)^2 + \frac{n(n-1)(n-2)}{3!}x^{n-3}$$

$$(\Delta x)^3 + \cdots + nx(\Delta x)^{n-1} + (\Delta x)^n$$

Subtracting x^n from both sides of the previous equation and dividing both sides of the equation by the increment of the variable x, and after separating in fractional terms and applying the Normalization factor of Newton, we found components of the Paraconsistent Newton's quotient, represented by the Equation (9), as shown below:

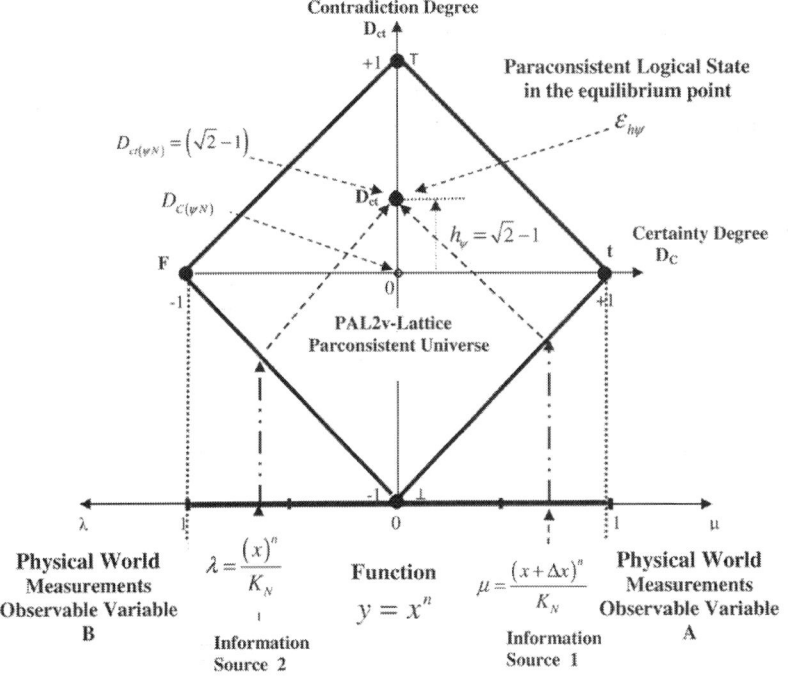

Figure 2: Location of the Paraconsistent logical state $\varepsilon_{h\psi}$ in the point at which Newton normalization factor (K_N) is used to adjust it at the equilibrium point $h_\psi = \sqrt{2} - 1$.

$$\frac{(x+\Delta x)^n}{K_N \Delta x} - \frac{(x)^n}{K_N \Delta x} = \frac{x^n}{K_N \Delta x} + \frac{nx^{n-1}(\Delta x)}{K_N \Delta x} + \frac{n(n-1)}{K_N \Delta x 2!}x^{n-2}(\Delta x)^2$$

$$+ \frac{n(n-1)(n-2)}{K_N \Delta x 3!}x^{n-3}(\Delta x)^3 + \cdots + \frac{nx(\Delta x)^{n-1}}{K_N \Delta x} + \frac{(\Delta x)^n}{K_N \Delta x} - \frac{x^n}{K_N \Delta x}$$

$$(31)$$

This equality Equation (31) compares Paraconsistent Newton's quotient with the Derivative equation of conventional method before the increase of variable x tends to zero. To make the increase of variable x tends to zerothe term fractional on the right side of the Equation (31) $\frac{x^n}{K_N \Delta x}$ is eliminated. This term, which is disallowed by the conventional method by applying the binomial theorem, is identified as the Unfavorable Evidence Degree $\lambda_{\psi N}$ of PAL2v analysis. Thereby the equation expresses the limit of the function, to be described as:

$$\lim_{\Delta x \to 0} \frac{\Delta y}{\Delta x} = nx^{n-1}$$

$$(32)$$

Or through another notation: $F'(x) = \lim\limits_{\Delta x \to 0} \frac{\Delta y}{\Delta x} = nx^{n-1}$ with the restriction that $x \neq 0$.

In conventional Integral method the equation of Primitive function should have adjusted their coefficient to adjust the values. Therefore, the Primitive Function of a Derivative function resulting $y' = x^n$ from conventional procedures will be: $F(x) = \frac{x^{n+1}}{(n+1)}$.

The value of the constant C is added to equation, thus obtaining the General Primitive function. And introducing the Indefinite Integral in symbolic mode, we have:

$$\int x^n dx = \frac{x^{n+1}}{(n+1)} + C$$

$$(33)$$

Equations of Paraconsistent Integral Calculus

In conventional Derivative [14] before considering action of x tend to zero; the application of binomial theorem allowed Primitive function of a Derivative function had potency n − 1, such that:

$F'(x) = nx^n \rightarrow F(x) = nx^{(n-1)}$.

In the Paraconsistent Logic this mathematical process indicates that in the derivative is performed a contraction in the PAL2v-Lattice. Other action of applying the binomial theorem is that when it is made the Derivative; the eliminated term is corresponding to the degree of Unfavorable Evidence Degree $\lambda_{\psi N}$ of the Paraconsistent Newton's quotient. For the Paraconsistent Logic this mathematical process that represents the action of Derivative modifies the Certainty Degree of the Primitive function. Thereby, the Paraconsistent Newton's quotient (Equation (14)) to the condition imposed by applying of the binomial theorem written in differential form will be:

$$PQ_{(I\psi N)}\Delta x = \left[\frac{(x+\Delta x)^n}{K_N} - 0 \right]$$

(34)

Therefore, for this condition, are identified:

$$\mu_{\psi N} = \frac{(x+\Delta x)^n}{K_N} \rightarrow \text{Favorable Evidence Degree and } \lambda_{\psi N} = 0 \rightarrow \text{Unfavor-}$$

able Evidence Degree.

Then, after the Derivative action, the Certainty degree, expressed by Equation (15) is:

$D_{C(I\psi N)} = \mu_{\psi N} - \lambda_{\psi N} \rightarrow D_{C(I\psi N)} = \mu_{\psi N} - 0$ resulting:

$$D_{C(I\psi N)} = \frac{(x+\Delta x)^n}{K_N}$$

(35)

Similarly, the Derivative action also modifies the value of the Contradiction Degree (Equation (16)), that before was expressed by:

$D_{Ct(I\psi N)} = \mu_{\psi N} - \lambda_{\psi N} - 1 \rightarrow D_{Ct(I\psi N)} = \mu_{\psi N} + 0 - 1$. Resulting:

$$D_{ct(I\psi N)} = \frac{(x + \Delta x)^n}{K_N} - 1$$

(36)

To happen the Derivative action that nullifies the Unfavorable Evidence Degree ($\lambda_{\psi N}$) and maintains the value of the Favorable Evidence Degree ($\mu_{\psi N}$), has an change in location of Paraconsistent logical state (ε) into PAL2v-Lattice. Figure 3 shows the location of Paraconsistent logical state (ε) after the derivative action of the Primitive function of Derivative function $y = nx^{n+1}$. The point where the Paraconsistent logical state (ε) will suffer the action of the integration process is called Paraconsistent logical state of Integral point. It is represented by $\varepsilon_{I\psi}$.

It is verified that the Integral Paraconsistent aims to return the Paraconsistent logical state (ε) to equilibrium point established by the Paraquantum Factor of quantization (h_ψ). Therefore, integral action will cause the Paraconsistent Logical State (ε), that after the Derivative action was located at the Integral point $\varepsilon_{I\psi}$ (shown in Figure 2), it is then restored to the equilibrium point of Paraconsistent Factor of quantization, represented by ($\varepsilon_{h\psi}$) in Figure 3. In this process of Integral Paraconsistent, which can be regarded as an anti-Derivative, when is added 1 to the n potency coefficient of x is promoted a first action in the PAL2v-Lattice expansion. Therefore, at the point of Integration $\varepsilon_{I\psi}$ the Favorable Evidence Degree is represented by: $\mu_{\psi N} = \frac{(x + \Delta x)^n}{K_N}$. With the expansion of the PAL2v-Lattice the Favorable Evidence Degree shall be represented by: $\mu_{\psi N} = \frac{(x + \Delta x)^{n+1}}{K_N(n+1)}$. The condition for Paraconsistent logical state (ε) is located at the Paraconsistent equilibrium point of quantization $\varepsilon_{h\psi}$ is that both Degrees of evidence should exist to form the Contradiction Degree of the Paraconsistent Newton's quotient. Therefore, as in the process of Derivative the Unfavorable Evidence Degree was made zero ($\lambda_{\psi N} = 0$), in the integral action it is reset for your expanded value that, in this process, establishes itself as:

$$\lambda_{\psi N} = \frac{(x)^{n+1}}{K_N(n+1)}.$$

At the equilibrium point, which is under the vertical axis of PAL2v-Lattice, the Contradiction Degree has its value known, such that:

$D_{Ct(\psi N)} = h_\psi = \sqrt{2} - 1$. Therefore, we can make the equality this known value, with the equation of Contradiction Degree where the Unfavorable Evidence Degree was reset:

$$D_{ct(\psi N)} = \left[\frac{(x+\Delta x)^{n+1}}{K_N(n+1)} + \frac{(x)^{n+1}}{K_N(n+1)} - 1 \right].$$

Resulting in:

$$D_{ct(\psi N)} + 1 = \frac{(x+\Delta x)^{n+1}}{K_N(n+1)} + \frac{(x)^{n+1}}{K_N(n+1)} = \sqrt{2}$$

Dividing the terms of equality in the previous equation:

$$\frac{D_{ct(\psi N)} + 1}{2} = \frac{(x+\Delta x)^{n+1}}{2K_N(n+1)} + \frac{(x)^{n+1}}{2K_N(n+1)} = \frac{\sqrt{2}}{2}.$$

Therefore, the value added to the Contradiction Degree of Paraconsistent Logic State at the integral point is

$\frac{\sqrt{2}}{2}$. This value obtained in the previous equality is enough for the Paraconsistent logical state (ε) reach the equilibrium point of Paraconsistent Factor of quantization $\varepsilon_{h\psi}$ in the expanded Lattice.

After the integral action the normalized Contradiction Degree presented in Equations (5) and (18) will be:

$$\mu_{ctr(\psi N)} = \frac{(x+\Delta x)^{n+1}}{2K_N(n+1)} + \frac{(x)^{n+1}}{2K_N(n+1)}$$

(37)

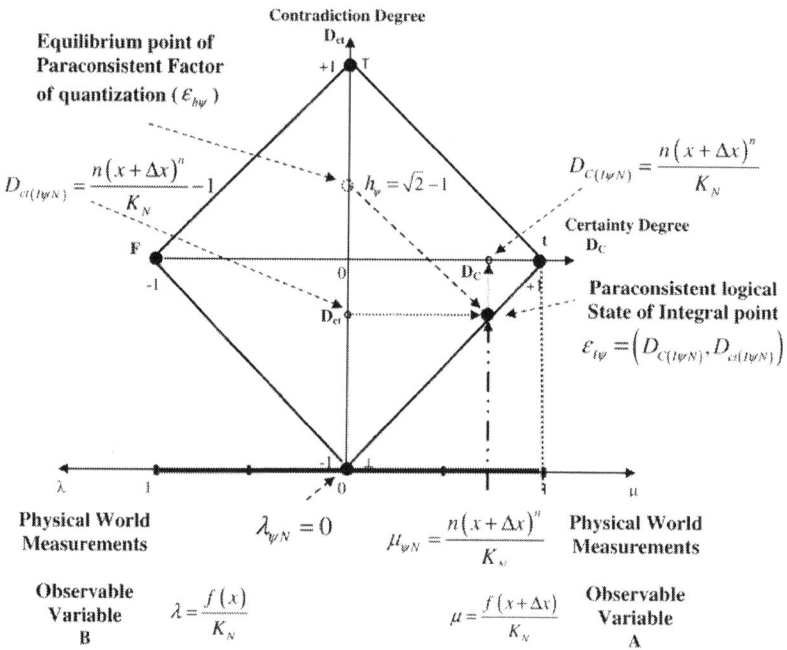

Figure 3: Location of Paraconsistent logical state of Integral point $\varepsilon_{I\psi}$ obtained after the action process of Derivative in a Primitive function of the Derivative function.

It is verified that Paraconsistent logical state from the equilibrium point of Paraconsistent Factor of quantization $\varepsilon_{h\psi}$ is Paraconsistent logical state of Primitive function, and it is located at the point of equilibrium determined by Paraconsistent Newton normalization Factor. Therefore the Primitive function in Paraconsistent Logical Model will be represented by the normalized Contradiction Degree of Equation (37), such that:

$$F_{(\psi N)} = \frac{(x + \Delta x)^{n+1}}{2K_N(n+1)} + \frac{(x)^{n+1}}{2K_N(n+1)}$$

(38)

where: K_N is a Normalization factor of Newton, such that: $K_N = \sqrt{2}y_{max}$

y_{max} is the maximum value of the function at the point considered.

Similarly, the value of the constant C is added to the Equation (38), thus obtaining the Primitive function General:

$$F(x)_N = \left[\frac{(x+\Delta x)^{n+1}}{2K_N(n+1)} + \frac{(x)^{n+1}}{2K_N(n+1)} \right] + C$$

(39)

Figure 4 shows the Paraconsistent integral action that is expanding PAL2v-Lattice. It is verified that the Paraconsistent Integral process takes the Contradiction Degree of Paraconsistent logical state of Integral point $\varepsilon_{I\psi}$ to the Paraconsistent logical state of equilibrium point of quantization $\varepsilon_{h\psi B}$.

Multiplies the value of K_N to the result obtained in the PAL2v-Lattice, and the Primitive function final will be given by:

$$F(x)_{\psi N} = \left[\frac{(x+\Delta x)^{n+1}}{2(n+1)} + \frac{(x)^{n+1}}{2(n+1)} \right] + C$$

(40)

The Paraconsistent Integral Undefined is presented in symbolic mode, such that:

$$\int x^n_{(\psi N)} dx = \left[\frac{(x+\Delta x)^{n+1}}{2(n+1)} + \frac{(x)^{n+1}}{2(n+1)} \right] + C$$

(41)

Thus, the calculation of the area will be:

1. For the area in the second point of the curve x = b: $A_b = \int x^n_{(\psi N)a} dx$

2. For the area at the first point of the curve x = a: $A_a = \int x^n_{(\psi N)a} dx$

The total area is calculated by:

$$A = \int x^n_{(\psi N)b} dx - \int x^n_{(\psi N)a} dx$$

(42)

APPLICATION EXAMPLES

Example 1

Consider as a first example that is given the Derivative function $F'(x) = x^2$ and we can find the Primitive function using the concepts of a Paraconsistent Logical Model.

Resolution: Initially, n appears in the Derivative function as: n = 2.

Using the Equation (39) the Primitive function by paraconsistent mode will be found:

$$F(x)_N = \left[\frac{(x+\Delta x)^{n+1}}{2(n+1)} + \frac{(x)^{n+1}}{2(n+1)} \right] \to F(x)_N = \left[\frac{(x+\Delta x)^{2+1}}{2(2+1)} + \frac{(x)^{2+1}}{2(2+1)} \right] \to$$

$$F(x)_N = \left[\frac{(x+\Delta x)^3}{6} + \frac{(x)^3}{6} \right]$$

Example 2

As second example considers that given a Derivative function from the type $F'(x) = 2x$, we wish to find the Primitive function.

Resolution: Note that the Derivative function n=1. Then using the Equation (41), the primitive function through the application of paraconsistent mode will be found, by:

$$F(x)_{\psi N} = \left[\frac{(x+\Delta x)^{n+1}}{2(n+1)} + \frac{(x+\Delta x)^{n+1}}{2(n+1)} \right] \to F(x)_{\psi N} = \left[\frac{2(x+\Delta x)^{1+1}}{2(1+1)} + \frac{2(x)^{1+1}}{2(1+1)} \right]$$

$$F(x)_{\psi N} = \left[\frac{2(x+\Delta x)^2}{4} + \frac{2(x)^2}{4} \right] \to F(x)_{\psi N} = \frac{(x+\Delta x)^2}{2} + \frac{(x)^2}{2}.$$

Example 3

Consider as a third example where we use a Paraconsistent Integral Calculus to determine the area under curve $y = x^2$, from point x = 1 at x = 2.

An Introduction to Paraconsistent Integral Differential Calculus

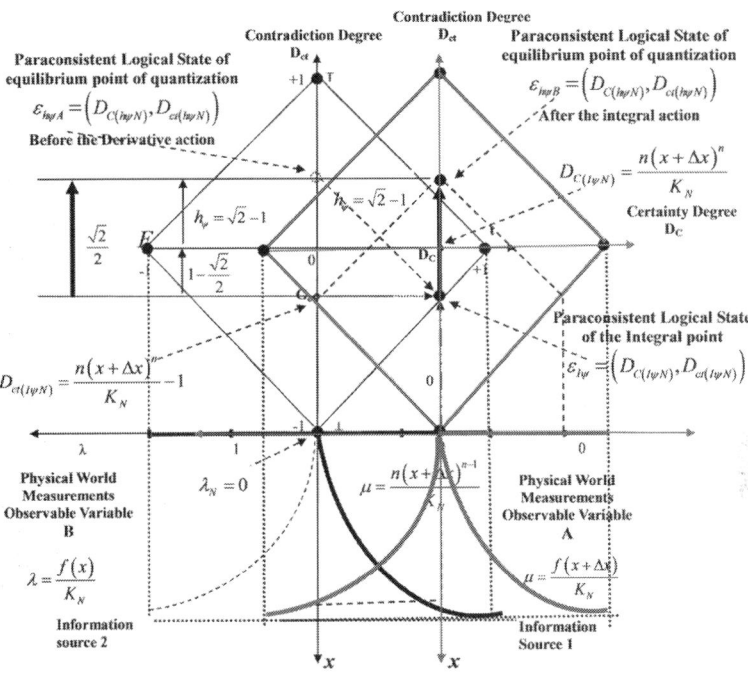

Figure 4: Final location of Paraconsistent logical state in a Paraconsistent integral process that expanding PAL2v-Lattice.

Resolution: In the resolution we can use the Equation (41) with an Increment value of variable x of $\Delta x = 0.001$:

$$\int x^n_{(\psi N)} \Delta x = \frac{(x + \Delta x)^{2+1}}{2(2+1)} + \frac{(x)^{2+1}}{2(2+1)} + C \rightarrow \int x^n_{(\psi N)} \Delta x = \frac{(x + \Delta x)^3}{6} + \frac{(x)^3}{6} + C$$

For $x_? = 2$:

$$\int x^n_{(\psi N)2} \Delta x = \frac{(2 + 0.001)^3}{6} + \frac{(2)^3}{6} + C \rightarrow \int x^n_{(\psi N)2} \Delta x = \frac{8.012006001}{6} + \frac{8.0}{6} + C.$$

Resulting:

$$\int x^n_{(\psi N)2} \Delta x = 2.668667667 + C.$$

For $x_1 = 1$:

$$\int x_{(\psi N)1}^n \Delta x = \frac{(1+0.001)^3}{6} + \frac{(1)^3}{6} + C \rightarrow \int x_{(\psi N)1}^n \Delta x = \frac{1.003003001}{6} + \frac{1.0}{6} + C$$

Resulting:

$$\int x_{(\psi N)1}^n \Delta x = 0.333833833 + C.$$

Area calculation by the equation (42):

$$A = \int x_{(\psi N)2}^n \Delta x - \int x_{(\psi N)1}^n \Delta x.$$

$$A = (2.668667667 + C) - (0.333833833 + C)$$

Resulting:

$$A = 2.334833834.$$

Example 4

Consider another example where is used the Paraconsistent Integral Calculus to determine the area under the curve y = 3x from x = 1 at x = 2.

Resolution: In the resolution, using the Equation (41) we can consider an Increment value of variable x: $\Delta x = 0.001$:

$$\int x_{(\psi N)}^n \Delta x = \frac{3(x+\Delta x)^{n+1}}{2(n+1)} + \frac{3(x)^{n+1}}{2(n+1)} + C$$

$$\int x_{(\psi N)}^n \Delta x = \frac{3(x+\Delta x)^{1+1}}{2(1+1)} + \frac{3(x)^{1+1}}{2(1+1)} + C \rightarrow \int x_{(\psi N)}^n \Delta x = \frac{3(x+\Delta x)^2}{4} + \frac{3(x)^2}{4} + C$$

For $x_2 = 2$:

$$\int x_{(\psi N)2}^n \Delta x = \frac{3(2+0.001)^2}{4} + \frac{3(2)^2}{4} + C \rightarrow \int x_{(\psi N)2}^n \Delta x = \frac{12.012003}{4} + \frac{12.00}{4} + C.$$

Resulting:

$$\int x_{(\psi N)2}^{n} \Delta x = 6.00300075 + C.$$

For $x_1 = 1$:

$$\int x_{(\psi N)1}^{n} \Delta x = \frac{3(1+0.001)^2}{4} + \frac{3(1)^2}{4} + C \rightarrow \int x_{(\psi N)1}^{n} \Delta x = \frac{3.006003}{4} + \frac{3}{4} + C.$$

Resulting:

$$\int x_{(\psi N)1}^{n} \Delta x = 1.50150075 + C.$$

With the subtraction of areas using Equation (42), we have:

$$A = (6.00300075 + C) - (1.50150075 + C) \rightarrow A = 4.5015.$$

Example 5

As the example 5 consider that using an increment value of the variable x of: $\Delta x = 0.001$, we wish to calculate the Paraconsistent Integral:

$$\int_1^2 {}_{\psi N}\left(x^2 + 3x\right) dx .$$

Resolution: The resolution is done using the Equation (41):

$$\int_1^2 x_{(\psi N)}^{n} \Delta x = \frac{(x+\Delta x)^{n+1}}{2(n+1)} + \frac{3(x+\Delta x)^{n+1}}{2(n+1)} + \frac{(x)^{n+1}}{2(n+1)} + \frac{3(x)^{n+1}}{2(n+1)} + C$$

$$\int_1^2 x_{(\psi N)}^{n} \Delta x = \frac{(x+\Delta x)^3}{6} + \frac{3(x+\Delta x)^2}{4} + \frac{(x)^3}{6} + \frac{3(x)^2}{4} + C$$

For $x_2 = 2$:

$$\int_1^2 x_{(\psi N)2}^{n} \Delta x = \frac{(2+0.001)^3}{6} + \frac{3(2+0.001)^2}{4} + \frac{(2)^3}{6} + \frac{3(2)^2}{4} + C$$

$$\int_1^2 x_{(\psi N)2}^{n} \Delta x = 2.668667667 + 6.00300075 + C \rightarrow \int_1^2 x_{(\psi N)2}^{n} \Delta x = 8.671668417 + C$$

For $x_1 = 1$:

$$\int x^n_{(\psi N)1} \Delta x = \frac{(1+0.001)^3}{6} + \frac{3(1+0.001)^2}{4} + \frac{(1)^3}{6} + \frac{3(1)^2}{4} + C$$

$$\int x^n_{(\psi N)1} \Delta x = 0.333833833 + 1.50150075 + C \rightarrow \int x^n_{(\psi N)1} \Delta x = 1.835334584 + C$$

We found the result of the area using the Equation (42):

$$\int_1^2 x^n_{(\psi N)} \Delta x = (8.671668417 + C) - (1.835334584 + C)$$

Resulting:

$$\int_1^2 x^n_{(\psi N)} \Delta x = 6.836333834.$$

CONCLUSIONS

This article presented a Paraconsistent Mathematics that structures a method for differential and Integral Calculus using the foundations of Paraconsistent Logic applied to Newton's quotient. The study allowed an adequacy of Differential Calculus to Paraconsistent logical model. With this, existing contradictions are accepted as inherent to a logical model based on real situations, therefore of an imperfect world. It was found that the Differential Calculus, structured in a Paraconsistent Logic that accepts contradictions, is able to dissolve the uncertainties, adding values that conventionally would be despised. Even requiring further testing involving more complex math functions the results obtained are very promising and suggest good perspectives for future applications of differential and Integral Paraconsistent Calculus.

REFERENCES

1. Da Costa, N.C.A. (2000) Paraconsistent Mathematics. In: Batens, D., Mortensen, C., Priest, G. and Bendegen van, J.P., Eds., I World Congress on Paraconsistency1998 Ghent, Belgium, Frontiers in Paraconsistent Logic: Proceedings, King's College Publications, London, 165-179.

2. Jas'kowski, S. (1969) Propositional Calculus for Contradictory Deductive Systems. Studia Logica, 24, 143-157. http://dx.doi.org/10.1007/BF02134311

3. Da Costa, N.C.A. (1986) On Paraconsistent Set Theory. Logique et Analyse, 115, 361-371.

4. Da Silva Filho, J.I., Lambert-Torres, G. and Abe, J.M. (2010) Uncertainty Treatment Using Paraconsistent Logic: Introducing Paraconsistent Artificial Neural Networks. IOS Press, Amsterdam, 328.

5. Da Silva Filho, J.I. (2011) Paraconsistent Annotated Logic in Analysis of Physical Systems: Introducing the Paraquantum $\gamma\psi$ Gamma Factor. Journal of Modern Physics, 2, 1455-1469. http://dx.doi.org/10.4236/jmp.2011.212180

6. Da Silva Filho, J.I. (2012) Analysis of the Emissions Spectral line of the Paraquantum with Hydrogen Atom. Journal of Modern Physics, 3, 233-254. http://dx.doi.org/10.4236/jmp.2012.33033

7. Da Silva Filho, J.I. (2012) An Introductory Study of the Hydrogen Atom with Paraquantum Logic. Journal of Modern Physics, 3, 312-333. http://dx.doi.org/10.4236/jmp.2012.34044

8. Da Silva Filho, J.I. (2011) Paraconsistent Annotated Logic in analysis of Physical Systems: Introducing the Paraquantum $h\psi$ Factor of Quantization. Journal of Modern Physics, 2, 1397-1409. http://dx.doi.org/10.4236/jmp.2011.211172

9. Stroyan, K.D. and Luxemburg, W.A.J. (1976) Introduction to the Theory of Infinitesimals. Academic Press, New York.

10. Bell, J.L. (1998) A Primer of Infinitesimal Analysis. Cambridge University Press, Cambridge.

11. Baron, M.E. (1969) The Origins of the Infinitesimal Calculus. Pergamon Press, Hungary.

12. Keisler, H.J. (1976) Elementary Calculus: An Infinitesimal Approach. 1st Edition, Prindle, Weber & Schmidt, Boston.

13. Diethelm, K. and Ford, N. (2004) Multi-Order Fractional Differential Equations and Their Numerical Solution. Applied Mathematics and Computation, 154, 621-640.http://dx.doi.org/10.1016/S0096-3003(03)00739-2

14. PI Tipler, A. and Llewellyn, R.A. (2007) Modern Physics. 5th Edition, W. H. Freeman and Company, New York.

15. Kleene, S.C. (1952) Introduction to Metamathematics. North Holland/Van Nostrand, Amsterdam/New York

CITATION

1. Da Silva Filho, J. I. (2014) An Introduction to Paraconsistent Integral Differential Calculus: With Application Examples. Applied Mathematics, 5, 949-962. doi: 10.4236/am.2014.56090.

Several New Types of Fixed Point Theorems and Their Applications to Two-Point Ordinary Differential Equations

Congjun Zhang[1], Jinlu Li[2], Yan Zhang[1], and Xiaoliang Feng[3]

[1]School of Applied Mathematics, Nanjing University of Finance and Economics, Nanjing, China
[2]Department of Mathematics, Shawnee State University, Portsmouth, USA
[3]Department of Mathematics, Nanjing University, Nanjing, China

ABSTRACT

The present paper is mainly concerned with several new types of fixed point theorems in different spaces such as cone metric spaces and fuzzy metric spaces. By using these obtained fixed point theorems, we then prove the existence and uniqueness of the solutions to two classes of two-point ordinary differential equation problems.

INTRODUCTION

The theory of the fixed point has important applications in fields such as differential equations, equilibrium problems, variational inequality, optimization problems, maxmin problems etc. (cf. Klaus Deimling [1], Congjun Zhang [2] for example), which has attracted many scholars' attention and became a hot topic in mathematics and applied mathematics field for a long time. In recent decades, many new types of fixed point theorems have been proposed (see [3-6] and the reference therein) and the generalizations of the existing ones have been dramatically developed in many ways. In [7], LongGuang Huang and Xian Zhang have introduced the notion of cone metric spaces and proved some fixed point theorems of contractive mappings on cone metric spaces. For fixed point theorems in fuzzy metric spaces, see

[8-12]. In [13-16], some scholars have proved the fixed point theorem in partial order metric space, and applied them to prove the existence and uniqueness of the solution to the two-point ordinary differential equation problems. Inspired by the recent progress in this fields, we will study in the present paper the existence and uniqueness of the fixed point for some special mappings in cone metric spaces and fuzzy metric spaces as well as their applications to the following two-point ordinary differential equations.

Problem (1):
$$\begin{cases} u'(t) = f\big(t, u(t)\big), & t \in [0, T]. \\ u(0) = u(T), \end{cases}$$

Where $T > 0$, $f : I \times R \rightarrow R$ is a continuous function satisfying some conditions which will be given explicitly later.

Problem (2):
$$\begin{cases} x'(t) = f\big(t, x\big) + g\big(t, x\big), & t \in [0, T] \\ x(0) = x_0, \end{cases}$$

Where, $T > 0$, $f : I \times R \rightarrow R$ is a continuous function satisfying some conditions which will be given explicitly later.

The paper is organized as follows. For the reader's convenience, we recall in Section 2 some definitions and lemmas in cone metric spaces and fuzzy metric spaces that will be used in the sequel. Section 3 is devoted to the investigation on the existence and uniqueness of the fixed point for some special mappings in cone metric spaces and fuzzy metric spaces. In last section, two-point ordinary differential equation problems are studied by using the results obtained in Section 3 and the existence and uniqueness of the solutions to such equations is established.

PRELIMINARIES AND ABSTRACT RESULTS

We recall in this section some definitions and lemmas in cone metric spaces and fuzzy metric spaces that will be used in the sequel.

Definition 1 [6]: Let (X, d) be a metric space and f a mapping from X to X. For any $z \in X$, define $f^0 z = z,$, $f^n z = f(f^{n-1}z)$ for $n \geq 1$. The sequence $\{f^n\}$ is called the orbit of f and f^n the n iterate of f.

Definition 2: A function $f : [0,\infty) \to [0,\infty)$ is called a ω-function if it is a monotone increasing function and satisfies that $f(0) = 0$ and for any $\varepsilon > 0$, there exists $M > 0$, such that $f(t) < \varepsilon$, for every $t \in (\varepsilon, M)$.

For example: $f(x) = (1/10)x$, defined on $[0,\infty)$, is a ω-function.

Definition 3 [7]: Let X be a nonempty set. Let E be a real Banach space, K a cone of E satisfying $\mathrm{int}\, k \neq \varnothing$, where $\mathrm{int}K$ denotes the interior of K. Define a partial order $\succeq$ on E based on K as follows: for any $x, y \in E$, , $y \succeq x$ if and only if $y - x \in K$, while $y \succ x$ means $y \succeq x$ and $y \neq x$, and $y \succ\succ x$ means $y - x \in \mathrm{int}K$. And the following convention is assumed: $x \preceq y$ if and only if $y \succeq x$, x=y if and only if $x \preceq y$ and $y \preceq x$.

If a mapping $d : X \times X \to E$ satisfies:

1. $d(x, y) \succeq 0$, for all x, $y \in X$, d(x, y)=0 if and only if x=y;

2. $d(x, y)=d(y, x)$, for all x, $y \in X$;

3. $d(x, y) \succeq d(x, z)+d(z, y)$, for all x, y, $z \in X$ then d is called a cone metric on X and (X, d) is called a cone metric space with respect to the Banach space E and the cone K in E.

Definition 4 [2]:
1) A cone K in a Banach space E is called normal, if there exists a number $M > 0$ such that for all x, y $\in$ K, $\theta \preceq x \preceq y$ implies $\|x\| \leq M\|y\|$,

where θ the zero element of the Banach space E. is The smallest M satisfying that inequality is denoted by M^*, and it is called the normal constant of K;

2) A cone K in a Banach space E is called regular if every increasing sequence which is bounded from above is convergent. That is, if $\{x_n\}$ is sequence such that $x_1 \leq x_2 \leq \cdots \leq x_n \leq \cdots \leq y$ for some $y \in E$, then there is $x \in E$ such that $\|x_n - x\| \to 0 (n \to \infty)$.

Remark 1:

1. For any normal cone K in a Banach space E, M^* exists and $M^* \geq 1$(see [2]); 2) Equivalently, a cone K is regular if and only if every decreasing sequence which is bounded from below is convergent. It is well known that a regular cone is a normal cone.

Definition 5 [7]: Let (X, d) be a cone metric space with respect to a Banach space E and a cone K in E. Let $\{x_n\}$ be a sequence in X (see [7]).

1. $\{x_n\}$is called a convergent sequence with limit x, if for any $c \in E$, there exists $n_0 \in N$ such that for every $n > n_0$, $d(x_n, x) \prec \prec c$ holds. In this case, we denote the limit of $\{x_n\}$ by $\lim_{n \to \infty} x_n = x$, or $x_n \to x(n \to \infty)$.

2. $\{x_n\}$ is called a Cauchy sequence on X, if for any $c \in E$ with $0 \prec c$, there exists $n_0 \in N$ such that for each m, $n > n_0$,$d(x_n, x_m) \prec \prec c$ holds.

3. We call X a complete cone metric space with respect to the Banach space E and the cone K in E, if every Cauchy sequence is convergent in X.

Remark 2: If K is a normal cone, then $\{x_n\}$ converges to x if and only if $d(x_n, x) \to \theta$, as $n \to \infty$ $\{x_n\}$ is a Cauchy sequence on X if and only if $d(x_n, x_m) \to \theta$ as m, $n \to \infty$ (see [7]).

Definition 6 [7]: Let (X, d) be a cone metric space with respect to a Banach space E and a cone K in E. If for any sequence $\{x_n\}$ in X, there exists a subsequence $\{x_{n_k}\}$ of $\{x_n\}$, such that $\{x_{n_k}\}$ is convergent in X. Then the cone metric space (X, d) is said to be sequentially compact.

Definition 7 [9, 10]: A binary operation $*:[0,1]^2 \to [0,1]$ is called a continuous t-norm, if the following conditions are satisfied:

1. $*$ is associative and commutative;

2. $*$ is continuous;

3. $a * 1 = a$ for all $a \in [0,1]$;

4. $a * b \leq c * d$ whenever $a \leq c$ and $b \leq d$,

for each $a,b,c,d \in [0,1]$. If it only satisfies conditions 1), 2) and 4), then it is called a t-norm.

Four typical examples of continuous t-norms are $a *_1 b = \min\{a,b\}$, $a *_2 b = \dfrac{ab}{\max\{a,b,\lambda\}}$ for $0 < \lambda < 1$ and $a *_3 b = ab, a *_4 b = \max\{a+b-1,0\}$.

Definition 8 [9, 10]: Let X be an arbitrary nonempty set. Let $*$ be a continuous t-norm and M a fuzzy set on $X^2 \times (0,\infty)$. If the following conditions satisfy:

1) $M(x, y, t) > 0$.

2) $M(x, y, t) = 1$ and only if $x=y$

3) $M(x, y, t) = M(y, x, t)$;

4) $M(x,y,t) * M(y,z,s) \leq M(x,z,t+s)$;

5) $M(x,y,.) : (0,\infty) \to [0,1]$ is continuous, for any $x,y,z \in X$ and t, s>0, then the 3-tuple (X, M, *) is called a fuzzy metric space.

Remark 3: For any $x,y,z \in X$, $M(x,y,.) : (0,\infty) \to [0,1]$ is a non-decreasing function (see [9, 10]).

Definition 9 [9, 10]: Let (X, M, *) be a fuzzy metric space and M a fuzzy set on $X^2 \times [0,\infty)$. M is said to satisfies the n-property on $X2 \times (0,\infty)$ if $\lim\limits_{n \to \infty} \left[M(x,y,k^n t) \right]^{n^p} = 1$ whenever $x,y \in X, k > 1$ and $p > 0$.

Definition 10: Let $(X, M, *)$ be a fuzzy metric space and M a fuzzy set on $X^2 \times [0,\infty)$. M is said to satisfies the τ-property on $X^2 \times (0,\infty)$ if $\lim_{n\to\infty}\left(x,y,k^n t\right) = 1$ for all $x,y \in X$ and $k>1$.

Definition 11 [11]: A function $f : R^+ \to R^+$ is said to satisfy ϕ-condition, if f is a strictly increasing function satisfying $f(0) = 0$ and $\lim_{n\to\infty} f^n(t) = \infty$ for any t>0, where $f^n(t) = f\left(f^{n-1}(t)\right)$.

Remark 4: If a function $f : R^+ \to R^+$ satisfies the ϕ-condition, then the following inequalities hold (see [11]):

1. $f(t)>t$, for all t>0;

2. $f^n(t)>f^{n-1}(t)> \quad >f(t)>t$, for each n=1, 2, and for all t>0.

Definition 12: Let $(X, M,*)$ be a fuzzy metric space, the fuzzy set M is said to have ϕ-property whenever $\lim_{n\to\infty} M(x,y,f^n t) = 1$ for all $x,y \in X$ where $f : R^+ \to R^+$ satisfying the ϕ-condition.

Definition 13 [9, 10]: Let $(X, M, *)$ be a fuzzy metric space and M a fuzzy set on $X^2 \times [0,\infty)$.

1. A sequence $\{X_n\}$ in X is said to be fuzzy-convergent to a point $x \in X$, if $\lim_{n\to\infty} M(x_n,x,t) = 1$ for all t>0.

2. A sequence $\{x_n\}$ in X is called a fuzzy-Cauchy sequence, if for each $0 < \varepsilon < 1$ and t>0, there exists $n_0 \in N$, such that $M(x_n,x_m,t) > 1-\varepsilon$ for each $m,n \geq n_0$.

3. A fuzzy metric space is called fuzzy-complete, if every fuzzy-Cauchy sequence is fuzzy-convergent.

Definition 14 [9,10]: Let $(X, M, *)$ be a fuzzy metric space. The fuzzy set M is said to be fuzzy-continuous on $X^2 \times (0,\infty)$, whenever any $\{(x_n,y_n,t_n)\}$ in $X^2 \times (0,\infty)$ which fuzzy-converges to $\{(x,y,t)\} \in X^2 \times (0,\infty)$ implies $\lim_{n\to\infty} M(x_n,y_n,t_n) = M(x,y,t)$.

Remark 5: M is a continuous function on $X^2 \times [0,\infty)$ (see [9, 10]).

Definition 15 [12]: Let (X, M, *) be a fuzzy metric space and M the fuzzy set on $X^2 \times [0,\infty)$. Denote by $H_0(X)$ the set of all compact subsets of X and define a function $H_m : H_0(X) \times H_0(X) \times (0,\infty) \to (0,\infty)$ by $H_m(A,B,t) = \min\{\inf_{a \in A} M(a,B,t), \inf_{b \in B} M(A,b,t)\}$ for any $A, B \in H_0(X)$ and any t>0, where $M(a,B,t) = \text{sub}_{b \in A} M(a,b,t)$ and $M(A,b,t) = \text{sub}_{b \in A} M(a,b,t)$.

Lemma 1 [6]: Let (X, d) be a complete metric space, $f : X \to X$, for the n iterate of f (n>1), the following statements hold:

1. If f^n has a unique fixed point, then f has a unique fixed point.

2. If there exists $z \in X$, such that the orbit of f^n converges to z, then the orbit of f converges to z.

3. If the orbit of f^n is a bounded sequence, then the orbit of f is a bounded sequence.

Lemma 2: Let (X, d) be a complete metric space and T an expansive and surjective mapping on X, then T has a unique fixed point.

Proof: We claim first that T is injective. To show this claim, assume, by the way of contradiction, that there exist $x_0 \neq y_0 \in X$ such that. $T(x_0) = T(y_0) \in X$ Since $x_0 \neq y_0 \in X$, then $d(x_0, y_0)>0$ holds. Since T is an expansive mapping, it implies $d(T(x_0),T(y_0))>0$. It contradicts to $T(x_0) = T(y_0)$, that is, $d(T(x_0),T(y_0))=0$, which implies T is a bijection. Hence T^{-1} exists and is a contraction mapping. By the contractive mapping priciple, there exists a unique $x_0 \in X$, such that $T^{-1}(x_0) = x_0 \in X$, that is $x_0 = T(x_0) \in X$. The proof is complete.

Lemma 3 [5]: Let (X, d) be a complete metric space and f a self-mapping on X. If the following condition satisfies, for any $\varepsilon > 0$, there exists $\delta > 0$, such that $\varepsilon < d(x,y) < \varepsilon + \delta$ implies $d(f(x),f(y)) < \varepsilon$, then f has a unique fixed point ξ on X, and $\lim_{n \to \infty} f^n(x) = \xi$ for any $x \in X$.

Lemma 4 [7]: Let (X, d) be a sequentially compact cone metric space with respect to a Banach space E and a regular cone K in E. Suppose a mapping $T : X \rightarrow X$ satisfies the contractive condition: $d(Tx, Ty) \prec d(x, y)$, for all $x, y \in X, x \neq y$, then T has a unique fixed point in X.

Lemma 5 [4]: Let (X, d) be a compact metric space and T a self-mapping on X. Assume that $12d(x, Tx) < d(x, y)$ implies $d(Tx, Ty) < d(x, y)$ for any $x, y \in X$, then T has a unique fixed point.

Lemma 6 [9, 10]: Let $(X, M, *)$ be a fuzzy metric space, $a * b \geq ab$ for all $a, b \in [0, 1]$ and M satisfy n- property. Let $\{x_n\}$ be a sequence in X such that for all $n \in N$, $M(s_n, x_{n+1}, kt) \geq M(x_{n-1}, x_n, t)$ for every $0 < k < 1$, then $\{x_n\}$ is a Cauchy sequence in X.

THE EXISTENCE THEOREM OF FIXED POINTS

In this section, we apply the concepts and lemmas provided in Section 2 to prove some existence theorems of fixed points for some mappings. These results will be used in the following section.

Theorem 1: Let (X, d) be a complete metric space and $f : X \rightarrow X$ a surjective mapping. If there exist $p \in N$ and $h > 1$ such that

$$d\left(f^p(x), f^p(y)\right) \geq hd(x, y)$$

(I)

holds for any $x, y \in X$, then there exists a unique fixed point of f.

Proof: For each $y \in f^2(X)$, since f is a surjective, then there exists $z \in f(x)$, such that f (z)=y, in the same way, there exist $x \in X$, such that f(x)=z, i.e. there exists $x \in X$, such that $f^2(x)=y$. We deduce by induction that $f^p(x)$ is also surjective, which combining (I) shows that $f^p(x)$ is an expansive mapping. By Lemma 2, there exists a unique fixed point of $f^p(x)$, then we know by Lemma 1 that there exists a unique fixed point of f. The proof is complete.

Remark 6: It is obvious that we can get Lemma 2 from Theorem 1. An example satisfying Theorem 1 is given below.

Example 1: Define $f : R \to R$ by

$$f(x) = \begin{cases} -\dfrac{x}{5}, & \text{if } x \le 0; \\ -10x, & \text{if } x > 0, \end{cases}$$

it is clear that f is a surjective self-mapping on R and $f^2(x) = 2x$. f^2 satisfies condition (I), i.e. p=2, then f has a fixed point, 0 is the fixed point in this example.

Theorem 2: Let (X, d) be a sequentially compact cone metric space with respect to a Banach space E and a normal cone K in E with normal constant M^*. Assume that T is a self-mapping on X and satisfies for any $x, y \in X$, implies $\dfrac{1}{2M^*}\|d(x, Tx)\| < \|d(x, y)\|$, then T has a unique fixed point.

Proof: We claim first that $\beta = 0$ where β is defined by

$$\beta = \inf_{x \in X} \|d(x, Tx)\|.$$

Using reduction to absurdity, we suppose $\beta > 0$. Since (X, d) is sequentially compact, we deduce from the definition of β that there exists a sequence $\{x_n\}$ such that

$$\|d(x_n, Tx_n)\| \to \beta \, (n \to \infty)$$

and

$$(x_n, Tx_n) \to (\xi, \eta)(n \to \infty)$$

for some $\xi, \eta \in X$. Observe that the normal constant $M^* \ge 1$, there exists $n_0 > 0$ such that for any $n \ge n_0$ the inequality

$$\frac{1}{2M^*}\left\|d\left(x_n, Tx_n\right)\right\| < \left\|d\left(x_n, \eta\right)\right\|$$

holds, which combining the given conditions shows that for any $n \geq n_0$,

$$\left\|d\left(Tx_n, T\eta\right)\right\| < \left\|d\left(x_n, \eta\right)\right\|.$$

By calculations we then have

$$\left\|d\left(T\eta, T^2\eta\right)\right\| < \left\|d\left(\eta, T\eta\right)\right\| = \lim_{n\to\infty}\left\|d\left(Tx_n, T\eta\right)\right\|$$

$$\leq \lim_{n\to\infty}\left\|d\left(x_n, \eta\right)\right\| = \beta$$

which contradicts to the definition of β.

We prove next that T has a fixed point. We proceed once more by using reduction to absurdity and suppose that T has no fixed point. Then for each $n \in N$,

$$0 < \frac{1}{2M^*}\left\|d\left(x_n, Tx_n\right)\right\| < \left\|d\left(x_n, Tx_n\right)\right\|,$$

which implies that for each $n \in N$,

$$\left\|d\left(Tx_n, T^2x_n\right)\right\| < \left\|d\left(x_n, Tx_n\right)\right\|.$$

By the triangle inequality in cone metric spaces, we have

$$d\left(\xi, T^2x_n\right) \preceq d\left(\xi, Tx_n\right) + d\left(Tx_n, T^2x_n\right),$$

then,

$$\lim_{n\to\infty}\left\|d\left(\xi, T^2x_n\right)\right\| \leq \lim_{n\to\infty}M^*\left\|\left(d\left(\xi, Tx_n\right) + d\left(Tx_n, T^2x_n\right)\right)\right\|$$

$$\leq \lim_{n\to\infty}M^*\left(\left\|\left(d\left(\xi, Tx_n\right)\right\| + \left\|d\left(x_n, Tx_n\right)\right)\right\|\right)$$

$$= M^*\left\|d\left(\xi, \eta\right) + d\left(\xi, \eta\right)\right\| = 0.$$

We claim that at least one of the following two inequalities should be hold:

$$\frac{1}{2M^*}\left\|d\left(Tx_n, T^2 x_n\right)\right\| < \left\|d\left(Tx_n, \xi\right)\right\|,$$

$$\frac{1}{2M^*}\left\|d\left(x_n, Tx_n\right)\right\| < \left\|d\left(x_n, \xi\right)\right\|,$$

otherwise, we reach a contradiction by the following calculations:

$$\left\|d\left(x_n, Tx_n\right)\right\| \le M^* \left\|\left(d\left(x_n, \xi\right)\right\| + \left\|d\left(Tx_n, \xi\right)\right)\right\|$$

$$\le M^* \left(\frac{1}{2M^*}\left\|d\left(Tx_n, T^2 x_n\right)\right\| + \frac{1}{2M^*}\left\|d\left(x_n, Tx_n\right)\right\|\right)$$

$$< \frac{1}{2}\left\|d\left(x_n, Tx_n\right)\right\| + \frac{1}{2}\left\|d\left(x_n, Tx_n\right)\right\| = \left\|d\left(x_n, Tx_n\right)\right\|.$$

If the first inequality of the above two holds, then

$$\left\|d\left(\xi, T\xi\right)\right\| \le M^* \left(\left\|d\left(\xi, T^2 x_n\right)\right\| + \left\|d\left(T\xi, T^2 x_n\right)\right\|\right)$$

$$\le M^* \left(\left\|d\left(\xi, T^2 x_n\right)\right\| + \left\|d\left(\xi, Tx_n\right)\right\|\right)$$

$$\to 0\left(n \to \infty\right),$$

if the other one holds, then

$$\left\|d\left(\xi, T\xi\right)\right\| \le M^* \left(\left\|d\left(\xi, Tx_n\right)\right\| + \left\|d\left(T\xi, Tx_n\right)\right\|\right)$$

$$\le M^* \left(\left\|d\left(\xi, Tx_n\right)\right\| + \left\|d\left(\xi, x_n\right)\right\|\right)]$$

$$\to 0\left(n \to \infty\right),$$

which show that $T\xi = \xi$ in each case, and the proof of the existence of the fixed point is complete.

We finally prove the uniqueness of the fixed point. Suppose $v \in X, v \ne u$ and $Tv = v$. Since, $\frac{1}{2N^*}\left\|d\left(u, Tu\right)\right\| = 0 < \left\|d\left(u, v\right)\right\|$ then $\left\|d\left(u, Tv\right)\right\| = \left\|d\left(Tu, Tv\right)\right\| < \left\|d\left(u, v\right)\right\|$, we reach a contradiction which completes the proof.

Remark 7: In [7], Long-Guang Huang and Xian Zhang have established a fixed point theorem in a sequentially compact cone metric space with respect to a Banach space E and a regular cone K in E (see Lemma 4), where the mapping $T : X \to X$ satisfies the contractive condition. In [4], Tomonari Suzuki has established a fixed point theorem in a compact metric space where the mapping T satisfying a condition similarly to condition (II) of theorem 2 (see Lemma 5). Observe that any regular cone is always normal, Theorem 2 is established under a different and weaker condition when comparing with Lemma 4 and generalize the results of Lemma 5 from compact metric spaces to sequentially compact cone metric spaces.

Theorem 3: Let (X, M, *) be a complete fuzzy metric space, where * is defined by $a * b = ab$ for any $a, b \in [0,1]$ and M a fuzzy set on $X^2 \times (0, \infty)$ satisfying ϕ-property. For a surjective function $f : X \to X$, if for any $x, y \in X, x \neq y$, the following inequality holds

$$M\big(f(x), f(y), t\big) \geq M\big(x, y, \phi(t)\big),$$

(II)

then f has a fixed point on X. If inequality (II) is strict, then f has a unique fixed point on X.

Proof: By choosing y=f(x), we deduce from (II) that for any $x \in X$,

$$M\big(f(x), f^2(x), t\big) \geq M\big(x, f(x), \phi(t)\big)$$

(III)

Proceed by introduction on n, we have for any $n \in N$

$$M\big(f^{n+1}(x), f^n(x), t\big) \geq M\big(f^n(x), f^{n-1}(x), \phi(t)\big).$$

For any n>m, we have

$$M\big(f^n(x), f^m(x), t\big)$$

$$\geq M\big(f^{n-1}(x), f^{m-1}(x), \phi(t)\big)$$

$$\geq M\big(f^{m-1}(x), f^{m-2}(x), \phi^2(t)\big) \geq \cdots$$

$$\geq M\big(f^{n-m-1}(x), f(x), \phi^{m-1}(t)\big)$$

$$\geq M\big(f^{n-m}(x), x, \phi^m(t)\big)$$

Observe that M satisfies ϕ-property, then

$$\lim_{m \to \infty} M\left(f^n(x), f^m(x), t\right)$$

$$\geq \lim_{m \to \infty} M\left(f^{n-m}(x), x, \phi^m(t)\right) = 1,$$

which shows that {$f^n(x)$} is a fuzzy-Cauchy sequence. Since (X, M, *) is complete, there exists $x_0 \in X$, such that

$$\lim_{n \to \infty} M\left(f^n(x), x_0, t\right) = 1.$$

Then by (II) and the nondecreasing property of M, we have

$$M\left(f^{n+1}(x), f(x_0), t\right) \geq M\left(f^n(x), x_0, \phi(t)\right)$$

$$\geq M\left(f^n(x), x_0, t\right),$$

for any $x, y \in X, x \neq y$. Since

$$\lim_{n \to \infty} M\left(f^n(x), x_0, t\right) = 1,$$

$$\lim_{n \to \infty} M\left(f^{n+1}(x), f(x_0), t\right) = 1.$$

We therefore deduce

$$\lim_{n \to \infty} f^n(x) = f(x_0) = x_0,$$

Which shows f has a fixed point on X.

If there exist $x_0, x_0' \in X, x_0 \neq x_0'$ such that $f(x_0) = x_0, f(x_0') = x_0'$, then by condition (II),

$$M\left(x_0, x_0', t\right) = M\left(f(x_0), f(x_0'), t\right)$$

$$> M\left(x_0, x_0', \phi(t)\right) \geq M\left(x_0, x_0', t\right).$$

It is a contradiction, hence $x_0 = x_0'$. We have now proved the uniqueness which complete the proof.

Corollary 1: Let $(X, M, *)$ be a complete fuzzy metric space and $f : X \rightarrow X$ a bijective mapping, where $*$ is defined by $a * b = ab$ for any $a, b \in [0, 1]$ and M a fuzzy set on $X^2 \times (0, \infty)$ satisfying ϕ-property. If for any $x, y \in X, x \neq y$,

$$M\left(f(x), f(y), \phi(t)\right) \leq M(x, y, t),$$

then f has a fixed point on X. If the above inequality is strict, then f has a unique fixed point on X.

Proof: Since f is bijective, $f^{-1} : X \rightarrow X$ exists and satisfies for any $x, y \in X, x \neq y$,

$$M\left(f^{-1}(x), f^{-1}(y), t\right) = M(x, y, t)$$
$$\geq M\left(f(x), f(y), \phi(t)\right).$$

By Theorem 3, we know f^{-1} has fixed point, and the fixed point of f^{-1} is the same as that of f, then f has fixed point on X. If the inequality is strict, then the proof is the same as that in Theorem 3.

Corollary 2: Let $(X, M, *)$ be a complete fuzzy metric space, where $*$ is defined by $a * b = ab$ for any $a, b \in [0, 1]$ and M a fuzzy set on $X^2 \times (0, \infty)$ satisfying $\tau -$ property, and $f : X \rightarrow X$ a surjective mapping satisfying

$$M\left(f(x), f(y), kt\right) \geq M(x, y, t),$$

for any $x, y \in X, x \neq y, , k \in (0, 1)$ Then f has a fixed point on X. If the inequality is strict, then f has a unique fixed point on X.

Proof: Let $\phi(t) = 1kt., 0 < k < 1$, then by Theorem 3 we can easily propose the results of Corollary 2. We omit the details.

Example 2: Assume $X = R.$, $t \in (0, \infty)$, and define M by

$$M(x, y, t) = e^{\frac{-d(x, y)}{t}},$$

clearly M satisfies $\tau-$ property. For any f satisfies the conditions of Corollary 2, i.e.

$$e^{\frac{-d(f(x),f(y))}{kt}} \geq e^{\frac{-d(x,y)}{t}},$$

we have $d(f(x),f(y)) \leq kd(x,y)$ for any $k \in (0,1)$, hence f is a contraction mapping which has a fixed point on X.

In the following, we show an example to demonstrate the conditions in Corollary 2 are only sufficient condition, not necessary conditions.

Example 3: Assume $X = R.$, $t \in (0,\infty)$, and define M by

$$M(x,y,t) = e^{d(x,y)t}.$$

Obviously M(x, y, t) is a fuzzy set which doesn't have $\tau-$ property, hence it can't be judged by Corollary 3. But if f is a contraction mapping, a fixed point still exist on X.

Theorem 4: Let (X, M, *) be a complete fuzzy metric space, where * is defined by $a*b = ab$ for any $a,b \in [0,1]$ and M a fuzzy set on $X^2 \times (0,\infty)$ satisfying ϕ-property. $F: X \to 2^X$ is a compact setvalued mapping, satisfies for any $x,y \in X$,

$$H_M(F(x),F(y),t) \geq M(x,y,\phi(t)),$$

then F has a fixed point on X.

Proof: By the choice axioms (see [6]), there exists a single-valued function $f: X \to X$, such that $f(x) \in F(x)$ for any $x \in X$. Then for each $x_1, y_1 \in X$, there exist $x_2 = f(x_1) \in F(x_1), y_2 = f(y_1) \in F(y_1)$. By the definition of H_M, we have

$$M(f(x_1),f(y_1),t) = M(x_2,y_2,t)$$
$$\geq H_M(F(x_1)F(y_1),t) \geq M(x_1,y_1,\phi(t)).$$

Theorem 3 shows that f has a fixed point $y_0 \in X$, i.e. $y_0 = f(y_0) \in F(y_0)$, which is also a fixed point of F on X.

Corollary 3: Let (X, M, *) be a complete fuzzy metric space and M a fuzzy set on $X^2 \times (0, \infty)$ satisfying $\tau -$ property. $F : X \to 2^X$ is a compact setvalued mapping satisfying for every $x, y \in X$,

$$H_M(F(x), F(y), kt) \geq M(x, y, t), 0 < k < 1.$$

Then F has a fixed point on X.

APPLICATIONS TO DIFFERENTIAL EQUATIONS

This section is concerned with the proof of the existence and uniqueness of the solutions to the two-point ordinary differential equations by using the fixed point theorems obtained in Section 3. The following are the main results.

Theorem 5: Assume that $f : I \times R \to R$ is a continuous function. If there exists $\lambda > 0$ such that the following inequalities

$$0 \leq f(t, y) + \lambda y - [f(t, x) + \lambda x] \leq \lambda \omega(y - x),$$

(IV)

hold for any $x, y \in R$ with $y \geq x$, where ω is an ω-function, then Problem(1) has a unique solution.

Proof: Problem (1) is equivalent to the integral equation

$$u(t) = \int_0^T G(t, s) \left[f(s, u(s)) + \lambda u(s) \right] ds,$$

where

$$G(t, s) = \begin{cases} \dfrac{e^{\lambda(T+s-t)}}{e^{\lambda T} - 1}, & 0 \leq s < t \leq T; \\[3mm] \dfrac{e^{\lambda(s-t)}}{e^{\lambda T} - 1}, & 0 \leq t < s \leq T. \end{cases}$$

Define

$$F:C\big([0,T],R\big)\to C\big([0,T],R\big),$$

by

$$(Fu)(t)=\int_0^T G(t,s)\Big[f\big(s,u(s)\big)+\lambda u(s)\Big]ds.$$

Note that if $u\in C\big([0,T],R\big)$ is a fixed point of F, then $u\in C\big([0,T],R\big)$ is a solution to Problem (1). Define a order relation in $C\big([0,T],R\big)$ by $u\leq v$ if and only if $u(t)\leq v(t)$ for every $t\in[0,T]$, for every $u,v\in C\big([0,T],R\big)$. Denote by $d(u,v)=\sup_{t\in[0,T]}\big|u(t)-v(t)\big|$ for any $u,v\in C\big([0,t],R\big)$ the distance in $C\big([0,T],R\big)$. For each $u>v$, by the left side of (IV), $f(t,u)+\lambda u\geq f(t,v)+\lambda v$. Since $G(t,s)>0$, for each $(t,s)\in[0,T]\times[0,T],t\geq s$,

$$(Fu)(t)=\int_0^T G(t,s)\Big[f\big(s,u(s)\big)+\lambda u(s)\Big]ds$$
$$\geq\int_0^T G(t,s)\Big[f\big(s,v(s)\big)+\lambda v(s)\Big]ds=(Fv)(t),$$

which shows that F is monotone increasing. For any $u\geq v$, if $\varepsilon<d(u,v)<\varepsilon+\delta$, then

$$d\big(Fu,Fv\big)\leq\sup_{t\in[0,T]}\big|(Fu)(t)-(Fv)(t)\big|$$

$$\leq\sup_{t\in[0,T]}\int_0^T G(t,s)$$

$$\cdot\big|f\big(s,u(s)\big)+\lambda u(s)-f\big(s,v(s)\big)-\lambda v(s)\big|ds$$

$$\leq\sup_{t\in[0,T]}\int_0^T G(t,s)\lambda\omega\big(u(s)-v(s)\big)ds$$

$$\leq\sup_{t\in[0,T]}\int_0^T G(t,s)\lambda\omega\big(d\big(u(s),v(s)\big)\big)ds.$$

Since $\omega(t)$ is a increasing function, then

$$\omega\big(d\big(u(s)\big)\big) \le \omega\big(d(u,v)\big) \text{ for } u>v, \text{ and}$$

$$d\big(Fu,Fv\big) \le \sup_{t\in[0,T]} \int_0^T G(t,s)\lambda\omega\big(d\big(u(s),v(s)\big)\big)ds$$

$$\le \lambda\omega\big(d(u,v)\big)\sup_{t\in[0,T]}\int_0^T G(t,s)ds$$

$$= \lambda\omega\big(d(u,v)\big)\cdot \sup_{t\in[0,T]}\frac{1}{e^{\lambda T}-1}\left(\frac{1}{\lambda}e^{\lambda(T+s-t)}\left|+\frac{1}{\lambda}e^{\lambda(s-t)}\right|\right)$$

$$\le \lambda\omega\big(d(u,v)\big)\frac{1}{\lambda\big(e^{\lambda T}-1\big)}\big(e^{\lambda T}-1\big)$$

$$\le \omega\big(d(u,v)\big).$$

By the definition of $\omega(t)$, for each $\varepsilon > 0$, there exists $M > 0$ such that $\varepsilon < d(u,v) < M, \omega\big(d(u,v)\big) < \varepsilon$, let $\delta = M - \varepsilon > 0$ hence $\varepsilon < d(u,v) < \varepsilon + \delta, \omega\big(d(u,v)\big) < \varepsilon$. It demonstrates $d(Fu,Fv,) < \varepsilon$. By Lemma 3, F has a unique fixed point, and $\lim_{n\to\infty} F^n(u_0) = u$ for each $u_0 \in C\big([0,T],R\big)$, u is the fixed point of F, i.e. the solution of Problem (1).

Assume $\alpha(t)$ is a lower solution of Problem (1), we can prove as Theorem 3.1 in [13] to obtain the uniqueness of the solution.

Remark 8: Contrasted with some related results in [13-15], the conditions in Theorem 5 is relatively clearer.

Theorem 6: Assume that $f : I \times R \to R$ is a continuous function. If there exists $\lambda > 0$ such that for any $x,y \in R$ with $y \ge x$, the following inequalities

$$0 \le f(t,y)-f(t,x)+g(t,y)-g(t,x) \le \lambda\eta(y-x) \tag{V}$$

hold, where η is an ω-function, then the solution of Problem (2) exists.

Proof: Problem (2) is equivalent to the following integral equation

$$x(t) = x_0 + \int_0^t f(s, x(s)) ds + \int_0^t g(s, x(s)) ds.$$

Define

$$F : C([0,T];R) \to C([0,T];R)$$

by

$$(Fu)(t) = x_0 + \int_0^t f(s, u(s)) + g(s, u(s)) ds,$$

for any $u(t,x) \in C([0,T];R)$. Note that $u \in C([0,T];R)$ is a fixed point of F, then $u \in C([0,T];R)$ is a solution of Problem (2). For $u, v \in C([0,T];R)$, we define $u \leq v$ if and only if $u(t) \leq v(t)$ for any $t \in [0,T]$. Denote $d(u,v) = \sup_{t \in [0,T]} |u(t) - v(t)|$, for. $u, v \in C([0,T];R)$ Then by (V), for any $u > v$,

$$f(t, u) \geq f(t, v),$$

$$f(t, u(s)) + g(t, u(s)) \geq f(t, v(s)) + g(t, v(s)),$$

which implies

$$(Fu)(t) = x_0 + \int_0^t f(s, u(s)) + g(s, u(s)) ds$$

$$\geq x_0 + \int_0^t f(s, v(s)) + g(s, v(s)) ds = (Fv)(t),$$

and

$$d(Fu, Fv) \le \sup_{t \in [0,T]} \left| (Fu)(t) - (Fv)(t) \right|$$

$$\le \sup_{t \in [0,T]} \left| \int_0^t \left[f(s, u(s)) - f(s, v(s)) \right] ds \right.$$

$$\left. + \int_0^t \left[g(s, u(s)) - g(s, v(s)) \right] ds \right|$$

$$\le \sup_{t \in [0,t]} \left| \int_0^T \lambda \eta(u(s) - v(s)) ds \right| \quad (0 < k < 1)$$

$$\le \sup_{t \in [0,T]} \left| \int_0^T \lambda \eta(d(u, v)) ds \right|.$$

By the definition of function ω, let $\varepsilon_1 = \varepsilon T \lambda > 0$, there exists $M > 0$, such that $\varepsilon_1 < d(u, v) < \varepsilon T \lambda, \omega(d(u, v)) < \varepsilon_1$, there exists $\delta > 0$, such that $M = \varepsilon + \delta$, then $\varepsilon_1 < d(u, v) < \varepsilon_1 + \delta$., $\eta(d(u, v)) < \varepsilon_1$ and

$$\sup_{t \in [0,T]} \left| \int_0^T \lambda \eta(d(u, v)) ds \right| \le \sup_{t \in [0,T]} \int_0^T \lambda \varepsilon ds \le 2T \lambda \varepsilon_1 = \varepsilon.$$

By Lemma 3, F has a unique fixed point, and $\lim_{n \to \infty} F^n(u_0) = u$ for any $u_0 \in C([0,T]; R)$, u is a fixed point of F, which is also a solution of Problem (2). The proof is complete.

Define $\theta : R \to R$ satisfying for any $u, v \in C([0,T]; R)$,

$$\theta(\lambda u(s)) - \theta(\lambda v(s))$$

$$\ge f(s, u(s)) - \lambda u(s) - (f(s, v(s)) - \lambda v(s))$$

then we have the following theorem:

Theorem 7: Let (X, M, *) be a complete fuzzy metric space,

$$M(x, y, t) = e^{\frac{-d(x,y)}{t}}.$$ If the following conditions hold:

1. For any $x, y \in R, x \ne y, \ k \in (0,1)$,

$$M(\theta(x), \theta(y), kt) \ge M(x, y, t);$$

2. For any $x \leq y$,

$$0 \leq f(t,y) + \lambda y - \left[f(t,x) + \lambda x \right],$$

then the solution of Problem (1) is unique.

Proof: By example 2, while $M(x,y,t) = e^{\frac{-d(x,y)}{t}}$, a mapping satisfying the above conditions is a contraction mapping, i.e. θ is a contraction mapping. Then we can proceed the proof with the same arguments as that in Theorem 5.

Remark 9: If we replace condition (1) by the inequality in Example 2 or Example 3 as well as the corresponding expression of M, then Theorem 7 can also make sure the uniqueness of the solution of Problem (1).

Define $h : R \to R$ satisfying for any $u, v \in C\left([0,T];R\right)$,

$$h\left(\lambda u(s)\right) - h\left(\lambda v(s)\right)$$
$$\geq \left[f\left(s, u(s)\right) - \left(f(s, v(s)) \right) \right]$$
$$+ \left[g\left(s, u(s)\right) - \left[g\left(s, v(s)\right) \right] \right],$$

then we have the following theorem:

Theorem 8: Let (X, M, *) be a complete fuzzy metric space, $M(x,y,t) = e^{\frac{-d(x,y)}{t}}$. If the following two conditions hold:

1. for any $x, y \in R, x \neq y$ and $k \in (0,1)$,,

$$M\left(h(x), h(y), kt\right) \geq M\left(x, y, t\right);$$

2. for any $x \leq y$,

$$0 \leq f(t,y) - f(t,x), 0 \leq g(t,y) - g(t,x),$$

then the solution of Problem (2) exists.

Proof: By Example 2, while $M(x,y,t) = e^{\frac{-d(x,y)}{t}}$, a mapping satisfying the conditions above is a contraction mapping, hence h is a contraction mapping. Then we can proceed the proof with the same arguments as that in Theorem 6 and complete the proof.

CONCLUSIONS

The paper is devoted to several new types of fixed point theorems in different spaces such as cone metric spaces and fuzzy metric spaces together with their applications. We have also proved the existence and uniqueness of the solutions to two classes of two-point ordinary differential equation problems by using these obtained fixed point theorems.

ACKNOWLEDGMENTS

The authors of this paper would like to appreciate the referee's helpful comments and valuable suggestions which have essentially improved this paper. This work is supported by the National Natural Science Foundation (11071109) of People's Republic of China.

REFERENCES

1. K. Deimling, "Nonlinear Functional Analysis," SpringerVerlag, Berlin, 1985. doi:10.1007/978-3-662-00547-7

2. C. J. Zhang, "Set-Valued Analysis and Its Applications to Economics," The Science Press, Beijing, 2004.

3. W. Walter, "Remarks on a Paper by F. Browder about Contraction," Nonlinear Analysis, Vol. 5, 1981, pp. 21-25. doi:10.1016/0362-546X(81)90066-3

4. T. Suzuki, "A New Type of Fixed Point Theorem in Metric Spaces," Nonlinear Analysis, Vol. 71, 2009, pp. 5313- 5317. doi:10.1016/j.na.2009.04.017

5. A. Meir and E. Keeler, "A Theorem on Contraction Mappings," Journal of Mathematical Analysis and Applications, Vol. 28, 1969, pp. 326-329. doi:10.1016/0022-247X(69)90031-6

6. T. Cardinali and P. Rubbioni, "An Extension to Multifunctions of the Keeler-Meir's Fixed Point Theorem," Fixed Point Theory, Vol. 7, No. 1, 2006, pp. 23-36.

7. L.-G. Huang and X. Zhang, "Cone Metric Spaces and Fixed Point Theorems of Contractive Mappings," Journal of Mathematical Analysis and Applications, Vol. 332, No. 2, 2007, pp. 1468-1476.
8. A. George and P. Veeramani, "On Some Result in Fuzzy Metric Space," Fuzzy Sets and Systems, Vol. 64, 1994, pp. 395-399. doi:10.1016/0165-0114(94)90162-7
9. S. Sedghi, I. Altunb and N. Shobe, "Coupled Fixed Point Theorems for Contractions in Fuzzy Metric Spaces," Nonlinear Analysis, Vol. 72, 2010, pp. 1298-1304. doi:10.1016/j.na.2009.08.018
10. X.-H. Zhu and J.-Z. Xiao, "Note on 'Coupled Fixed Point Theorems for Contractions in Fuzzy Metric Spaces'," Nonlinear Analysis, Vol. 74, 2011, pp. 5475-5479.doi:10.1016/j.na.2011.05.034
11. S. S. Zhang, "Fixed Point Theorems of Mappings on Probabilistic Metric Spaces with Applications," Scientia Sinca (Series A), Vol. 11, 1983.
12. J. Rodriguez-Lopez and S. Romaguera, "The Hausdorff Fuzzy Metric on Compact Sets," Fuzzy Sets and Systems, Vol. 147, No. 2, 2004, pp. 273-283.
13. A. Amini-Harandi and H. Emami, "A Fixed Point Theorem for Contraction Type Maps in Partially Ordered Metric Spaces and Application to Ordinary Differential Equations," Nonlinear Analysis, Vol. 72, 2010, pp. 2238-2242. doi:10.1016/j.na.2009.10.023
14. T. G. Bhaskar and V. Lakshmikantham, "Fixed Point Theorems in Partially Ordered Metric Spaces and Applications," Nonlinear Analysis, Vol. 65, 2006, pp. 1379-1393.doi:10.1016/j.na.2005.10.017
15. J. Harjani and K. Sadarangani, "Generalized Contractions in Partially Ordered Metric Spaces and Applications to Ordinary Differential Equations," Nonlinear Analysis, Vol. 72, 2010, pp. 1188-1197. doi:10.1016/j.na.2009.08.003
16. J. Harjani and K. Sadarangani, "Fixed Point Theorems for Weakly Contractive Mappings in Partially Ordered Sets," Nonlinear Analysis, Vol. 71, 2009, pp. 3403-3410.doi:10.1016/j.na.2009.01.240

CITATION

1. C. Zhang, J. Li, Y. Zhang and X. Feng, "Several New Types of Fixed Point Theorems and Their Applications to Two-Point Ordinary Differential Equations," Applied Mathematics, Vol. 3 No. 10, 2012, pp. 1109-1116. doi: 10.4236/am.2012.310163.

Differential Calculus for Some p-norms of the Fundamental Matrix with Applications

L. Kohaupt

Prager Str. 9, D-10779 Berlin, Germany

ABSTRACT

For the fundamental matrix $\Phi(t) = e^{At}$ of a complex n×n matrix A, the differential properties of the mapping $t \mapsto \|\Phi(t)\|_p$ at every point $t = t_0 \in \mathbb{R}_0^+ := \{t \in \mathbb{R} \mid t \geq 0\}$ are investigated, where $\|\cdot\|_p$ is the matrix operator norm associated with the vector norm $\|\cdot\|p$ in $\mathbb{C}^n$ or $\mathbb{R}^n$ as the case may be, for $p \in \{1, 2, \infty\}$. Moreover, formulae for the first two right derivatives $D_+^k \|\Phi(t)\|_p, k = 1, 2$, are calculated and applied to determine the best upper bounds on $\|\Phi(t)\| p$ in certain classes of bounds. These results cannot be obtained by the methods used so far. The systematic use of the differential calculus for norms, as done here for the first time, could lead to major advances also in other branches of mathematics and of other sciences, notably in engineering, for example in the simulation of dynamic problems with excitation.

INTRODUCTION

The purpose of this paper is twofold. First, it rounds off and completes results of [7] by developing a systematic study of the differential properties for some p-norms of the fundamental matrix, that is, a pertinent differential calculus. Second, the results of [7] and of this paper are applied to determine upper bounds on the norm of the fundamental matrix for a vibration problem; the obtained upper bounds are the best possible ones in the considered classes of bounds, a result which in general cannot be achieved by the methods used so far.

Starting point of the paper [7] was the notion of logarithmic derivative $\mu[A]$ of a complex $n \times n$ matrix A, which had been introduced in [8] in the context of error estimates for the numerical integration of ordinary differential equations. The logarithmic derivative is defined as the (first) right derivative of the function $t \mapsto \|\Phi(t)\|$ at t=0 where $\|\cdot\|$ is the matrix operator norm pertinent to a vector norm $\|\cdot\|$ in $\mathbb{C}^n$ or $\mathbb{R}^n$ as the case may be. The existence proof for $\mu[A]$ exploits that $\|\cdot\|$ is a convex function. In [8], also formulae for the logarithmic derivatives in the p-norms, $\mu_p[A], p \in \{1, 2, \infty\}$, are derived. The concept of logarithmic derivative $\mu[A]$ is used in [2] and [1] in the theory of ordinary differential equations to obtain new results, e.g., in stability problems, and the results improve those obtained by using the norm $\|A\|$. For example, one gets the general estimate $\|\Phi(t)\| \le e^{\mu[A]t}, t \ge 0$, which is better than $\|\Phi(t)\| \le e^{\|A\|t}, t \ge 0$, particularly, when μ [A] is negative. But, even though the concept of logarithmic derivative has brought considerable progress, it does not always deliver the optimal result. Even worse, in some cases, it gives useless results. For instance, in the applications at the end of this paper, we shall see that the estimate $\|\Phi(t)\|_p \le e^{\mu_p[A]t}, t \ge 0$ for the considered system matrix A is practically worthless for large $t \ge 0$ since it is by far too large. The reason for this is very simple. Because $\mu[A]$ describes the change of $\|\Phi(t)\|$ at the origin, one cannot expect that $\mu[A]$ furnishes good results for large $t \ge 0$. Apart from [2] and [1], many publications on $\mu[A]$ have appeared. In [3], $\mu[A]$ is called matrix measure.

Now, in [7] — more than 40 years after the definition of $\mu[A]$ — the author has gone further by showing that the mapping $t \mapsto \|\Phi(t)\|_\infty$ is real analytic in some neighborhood $[0, t_0]$, where $t_0 > 0$. Of fundamental importance in this context were two points (which turn out to be indispensable in this paper, too)

- The series expansion of $\|\Phi(t)\|_\infty$ and
- Lemma 2.1 in [7].

Moreover, for complex matrices A, a formula for the second logarithmic derivative $\mu_\infty^{(2)}[A]$ was derived and, for real matrices A, unified formulae for all logarithmic derivatives $\mu_\infty^{(k)}[A], k=1,2\ldots\ldots,.$ are obtained in [7]. The results were illustrated by an example, but applications were not yet given. It was mentioned there that the case p=1 can be treated by interchanging the row index and the column index of the matrix A in the formula for $\mu_\infty^{(k)}[A], k=1,2\ldots\ldots,..$ The case p=2 had not yet been resolved.

In this paper, we tackle the case p=2 and show that the second logarithmic derivative $\mu_2^{(2)}[A]$ is always nonnegative. This is surprising since $\mu_\infty^{(2)}[A]$ may be negative, zero, or positive. The fact that $\mu_2^{(2)}[A] \geqslant 0$ has far-reaching consequences as we shall see in the application part. The formulae for $\mu_p^{(k)}[A], p \in \{\infty, 2\}, k \in \{1,2\}$, are applied by deriving new upper bounds on $\|\Phi(t)\|_p$ in a neighborhood $\left[0, t_p^*\right]$ of t=0 (i.e., "near the origin"), which are better than $e^{\mu[A]t}$. In order to obtain ameliorations in the adjacent interval $\left[t_p^*, \infty\right]$ (which is "far from the origin") one has to carry over the differentiability results for the mapping $t \mapsto \|\Phi(t)\|_p$ at $t_0=0$ to every $t_0>0$. Since this offers no difficulty for the case p=∞, we state the results without proof. For the case p=2, the results follow directly from general theorems in [6] and [7, Lemma 2.1]. For the sake of completeness and ease of reference, the results are stated, and short proofs are given. For, it would be unreasonable to leave this to the individual interested readers.

Since on every finite interval of $\mathbb{R}_0^+$ the function $t \mapsto \|\Phi(t)\|_p$ is infinitely often differentiable with the possible exception of a finite number of points, we have almost everywhere on $\mathbb{R}_0^+$ the relations $D^k\|\Phi(t)\|_p = D_+^k\|\Phi(t)\|_p, k=1,2,\ldots\ldots; p \in \{1,2,\infty\}$. In this case, the differential calculus can be applied. We do this by determining the best constant $M_{\varepsilon,p}$ in the well-known upper bounds $\|\Phi(t)\|_p \leq M_{\varepsilon,p}e^{(\alpha+\varepsilon)t}$ for the differential equation $\dot{x} = Ax$, where α is the spectral abscissa of the matrix A and where $\varepsilon>0$ is any given number; we also compare the results with those obtained by the method used so far.

The actual new point of this paper is this systematic use of the differential calculus to obtain better results, and not so much the new lemmata and theorems.

The paper is structured as follows. In Section 2, regularity properties of the mapping $t \mapsto \|\Phi(t)\|_p$, $p \in \{\infty, 2\}$, at every point of $t = t_0 = \mathbb{R}_0^+$ are stated. In Section 3, formulae for the right derivatives can be found. Section 4 is the application part. Here, the results from the previous sections are applied to a vibration problem to obtain the best upper bounds in certain classes of bounds. Further, it is pointed out that the developed differential calculus can be used for the discussion of the functions $t \mapsto \|\Phi(t)\|_p$, $p \in \{1, 2, \infty\}$, e.g., for the determination of relative extrema and inflexion points. The case p=1 is not treated since the associated results can easily be obtained from the case p=∞. Finally, inSection 5 some concluding remarks are made.

LOCAL REGULARITY

We have the following lemma, which states — loosely speaking — that for every $t_0 \geq 0$ and for $p \in \{\infty, 2\}$ the function $t \mapsto \|\Phi(t)\|p$ is real analytic in some neighborhood $[t_0, t_0+\Delta t_0]$.

Lemma 1: Let$p \in \{\infty, 2\}$ and $t_0 \in \mathbb{R}_0^+$. Then, there exists a number $\Delta t_0 > 0$ and a function $t \mapsto \hat{\Phi}(t)$ which is real analytic on $[t_0, t_0+\Delta t_0]$, such that $\hat{\Phi}(t) = \|\Phi(t)\|_p$ for all $t \in [t_0, t_0+\Delta t_0]$.

Proof:

i.   Case$p=\infty$. The proof is similar to that for $t_0=0$ in [7].
ii. Case$p=2$. Let

$$\Psi(t) := \Phi^*(t)\,\Phi(t), \quad t \geq 0 \tag{1}$$

and let

$$\hat{\Psi} : z \mapsto \hat{\Psi}(z), \quad z \in \mathbb{C} \tag{2}$$

be the unique extension of $t \mapsto \Psi(t), t \geqslant 0$. Further, let $\lambda_j(\hat{\Psi}(z))$ be the eigenvalues of $\hat{\Psi}(z), z \in \mathbb{C}, j = 1,....,n$. Then, the functions

$$z \mapsto \lambda_j(\hat{\Psi}(z)), \quad z \in \mathbb{C} \tag{3}$$

j=1,...,n are holomorphic at every point of the real line. This is seen as follows: $\hat{\Psi}(\cdot)$ is Hermitian and thus normal at every point of the real line. Therefore, the conditions of [6, Chapter II, Theorem 1.10, p. 71] are fulfilled if the domain D_0 there is taken as the z-plane (cf. [6, p. 63] for D_0).

Let $\lambda_j(\Psi(t)), j = 1,.....,n$ be the eigenvalues of $\Psi(t), t \geqslant 0$. Then, the functions

$$t \mapsto \lambda_j(\Psi(t)), \quad t \geqslant 0 \tag{4}$$

are real analytic due to what has been said before. Consequently, for every $t_0 > 0$ there exists a number $\Delta t_0 > 0$ such that

$$\varphi_i(t) := [\lambda_i(\Psi(t))]^{1/2} = \sum_{k=0}^{\infty} \varphi_i^{(k)} \frac{(t - t_0)^k}{k!}, \quad t_0 \leqslant t \leqslant t_0 + \Delta t_0 \tag{5}$$

with real coefficients $\varphi_i^{(k)}, k = 0, 1, 2, \cdots; i = 1, \cdots, n$. Then,

$$\|\Phi(t)\|_2 = \max_{i=1,...,n} \varphi_i(t), \quad t_0 \leqslant t \leqslant t_0 + \Delta t_0 \tag{6}$$

The rest of the proof is similar to that of [7, Lemma 2.1].

FORMULAE FOR THE RIGHT DERIVATIVES

The formulae for the case p=∞ are stated without proof. Those for the case p=2 are stated for ease of reference and are proved for the sake of completeness.

Case p=∞: Complex n×n matrix A

Let $t_0 \in \mathbb{R}_0^+ = \{t \in \mathbb{R} \mid t \geq 0\}$, and for i,j=1,...,n define the functionals

$$\lambda_{ij}^{(0)}[A, t_0] := |\Phi_{ij}(t_0)| \tag{7}$$

$$\lambda_{ij}^{(1)}[A, t_0] := \begin{cases} \dfrac{\operatorname{Re}\Phi_{ij}(t_0)\operatorname{Re}(A\,\Phi)_{ij}(t_0) + \operatorname{Im}\Phi_{ij}(t_0)\operatorname{Im}(A\,\Phi)_{ij}(t_0)}{|\Phi_{ij}(t_0)|}, & \Phi_{ij}(t_0) \neq 0, \\[2mm] |(A\,\Phi)_{ij}(t_0)|, & \Phi_{ij}(t_0) = 0, \end{cases} \tag{8}$$

$\lambda_{ij}^{(2)}[A, t_0] :=$

$$\begin{cases} \dfrac{|(A\,\Phi)_{ij}(t_0)|^2 + \operatorname{Re}\Phi_{ij}(t_0)\operatorname{Re}(A^2\,\Phi)_{ij}(t_0) + \operatorname{Im}\Phi_{ij}(t_0)\operatorname{Im}(A^2\,\Phi)_{ij}(t_0)}{|\Phi_{ij}(t_0)|} \\[2mm] \quad - \dfrac{[\operatorname{Re}\Phi_{ij}(t_0)\operatorname{Re}(A\,\Phi)_{ij}(t_0) + \operatorname{Im}\Phi_{ij}(t_0)\operatorname{Im}(A\,\Phi)_{ij}(t_0)]^2}{|\Phi_{ij}(t_0)|^3}, & \Phi_{ij}(t_0) \neq 0 \\[3mm] \dfrac{\operatorname{Re}(A\,\Phi)_{ij}(t_0)\operatorname{Re}(A^2\,\Phi)_{ij}(t_0) + \operatorname{Im}(A\,\Phi)_{ij}(t_0)\operatorname{Im}(A^2\,\Phi)_{ij}(t_0)}{|(A\,\Phi)_{ij}(t_0)|}, & \Phi_{ij}(t_0) = 0,\ (A\,\Phi)_{ij}(t_0) \neq 0, \\[3mm] |(A^2\,\Phi)_{ij}(t_0)|, & \Phi_{ij}(t_0) = 0,\ (A\,\Phi)_{ij}(t_0) = 0, \end{cases} \tag{9}$$

where $(A\Phi)_{ij}(t_0) := [(A\Phi)(t_0)]_{ij}$, and so on. Let

$$\lambda_i^{(k)}[A, t_0] := \sum_{j=1}^{n} \lambda_{ij}^{(k)}[A, t_0], \quad i = 1,\ldots,n;\ k = 0,1,2 \tag{10}$$

Then, we obtain the following theorem.

Theorem 2: Let $A \in \mathbb{C}^{n\times n}, I_{-1} := \{1,\ldots,n\}$ and I_0 be the index set where $\lambda_i^{(0)}[A, t_0]$ attains its maximum,

$$I_0 := \left\{ i_0 \in I_{-1} \mid \lambda_{i_0}^{(0)}[A, t_0] = \max_{i \in I_{-1}} \lambda_i^{(0)}[A, t_0] \right\} \tag{11}$$

Similarly, let

$$I_1 := \left\{ i_1 \in I_0 \mid \lambda_{i_1}^{(1)}[A, t_0] = \max_{i \in I_0} \lambda_i^{(1)}[A, t_0] \right\} \tag{12}$$

and

$$I_2 := \left\{ i_2 \in I_1 \mid \lambda_{i_2}^{(2)}[A, t_0] = \max_{i \in I_1} \lambda_i^{(2)}[A, t_0] \right\} \tag{13}$$

Then,

$$\|\Phi(t_0)\|_\infty = \max_{i \in I_{-1}} \lambda_i^{(0)}[A, t_0] \tag{14}$$

$$D_+^1 \|\Phi(t_0)\|_\infty = \max_{i \in I_0} \lambda_i^{(1)}[A, t_0] \tag{15}$$

$$D_+^2 \|\Phi(t_0)\|_\infty = \max_{i \in I_1} \lambda_i^{(2)}[A, t_0] \tag{16}$$

Case p=∞: Real n×n matrix A

Define the following sign functionals

$$s_{ij}^{(0)}[\Phi(t_0)] := \mathrm{sgn}[\Phi_{ij}(t_0)] \tag{17}$$

and

$$\Phi(t_0)] := \begin{cases} \mathrm{sgn}[\Phi_{ij}(t_0)], & \Phi_{ij}(t_0) \neq 0, \\ \mathrm{sgn}[(A\,\Phi)_{ij}(t_0)], & \Phi_{ij}(t_0) = 0, (A\,\Phi)_{ij}(t_0) \neq 0, \\ \mathrm{sgn}[(A^2\,\Phi)_{ij}(t_0)], & \Phi_{ij}(t_0) = 0, (A\,\Phi)_{ij}(t_0) = 0, (A^2\,\Phi)_{ij}(t_0) \neq 0 \\ \vdots \\ \mathrm{sgn}[(A^k\,\Phi)_{ij}(t_0)], & (A^l\,\Phi)_{ij}(t_0) = 0, \ l = 0,1,\ldots,k-1, \end{cases} \tag{18}$$

i,j=1,..., n; k=1,2,.... This relation can also be written as

$$s_{ij}^{(k)}[A^k\,\Phi(t_0)] = \begin{cases} s_{ij}^{(k-1)}[A^{k-1}\,\Phi(t_0)], & s_{ij}^{(k-1)}[A^{k-1}\,\Phi(t_0)] \neq 0, \\ \mathrm{sgn}[(A^k\,\Phi)_{ij}(t_0)], & s_{ij}^{(k-1)}[A^{k-1}\,\Phi(t_0)] = 0, \end{cases} \tag{19}$$

k=1,2,.... With these sign functionals, define the further functionals

$$\lambda_i^{(k)}[A, t_0] := \sum_{j=1}^n s_{ij}^{(k)}[A^k\,\Phi(t_0)](A^k\,\Phi)_{ij}(t_0) \tag{20}$$

$i=1,...,n; k=1,2,....$ Then, the right derivatives for real matrices read as follows.

Theorem 3: Let $A \in \mathbb{R}^{n \times n}$ define $I_{-1}=\{1,...,n\}$ and let I_k be the set of all indices $i_k \in I_{k-1}$, where $\lambda_i^{(k)}[A,t_0]$ from (20) attains its maximum, i.e. $k=0,1,2,....$ Then, the right derivatives of $t \mapsto \|\Phi(t)\|_\infty$ at $t=t_0>0$ are given by

$$D_+^k \|\Phi(t_0)\|_\infty = \max_{i \in I_{k-1}} \lambda_i^{(k)}[A,t_0]$$

(22)

$k=0,1,2,....$

Case p=2: Real or complex $n \times n$ matrix A

Starting point in this case is the series expansion

$$\Psi(t) := \Phi^*(t)\,\Phi(t) = \sum_{j=0}^{\infty} \Phi^*(t_0)\,B_j\,\Phi(t_0)\,\frac{(t-t_0)^j}{j!}$$

(23)

with

$$B_j = \sum_{k=0}^{j} \binom{j}{k} A^{*j-k} A^k,$$

(24)

$j=0,1,2,....$ Thus, e.g.,

$$B_0 = E,$$

(25)

$$B_1 = A^* + A,$$

(26)

$$B_2 = A^{*2} + 2A^* A + A^2$$

(27)

Consequently,

$$\Psi(t) = T^{(0)} + T^{(1)}(t-t_0) + T^{(2)}(t-t_0)^2 + \cdot$$

(28)

with

$$T^{(0)} = \Phi^*(t_0)\,\Phi(t_0),$$

(29)

$$T^{(1)} = \Phi^*(t_0) B_1 \Phi(t_0),$$ (30)

$$T^{(2)} = \Phi^*(t_0)(\tfrac{1}{2} B_2)\Phi(t_0).$$ (31)

Let $\lambda_{max}(\Psi(t))$ be the largest eigenvalue of $\Psi(t)$. Then, due to [6, Theorem 5.11, Chapter II, pp. 115–116; 7, Lemma 2.1]

$$\lambda_{max}(\Psi(t)) = v_0 + v_1(t - t_0) + v_2(t - t_0)^2 + \cdots, \quad t_0 \leqslant t \leqslant t_0 + \Delta t_0$$ (32)

where the quantities v_0, v_1, and v_2 are derived now.

Let $n_{-1} := n$ and $v_k^{(0)}[T^{(0)}], k = 1,....,n_{-1}$, be the eigenvalues of $T^{(0)}$. Then,

$$v_0 = \max_{k=1,...,n_{-1}} v_k^{(0)}[T^{(0)}] = \|\Phi(t_0)\|_2^2$$ (33)

Further, define

$$M_{-1} := X := \mathbb{C}^n$$

Let

$$V_0 := [v_1^{(v_0)},\ldots, v_{n_0}^{(v_0)}]$$

be the matrix formed by the orthonormal set of eigenvectors $v_k^{(v_0)}, k = 1,....,n_0$ associated with V_0, and let P_{V_0} be the orthogonal projection on the algebraic eigenspace

$$M_0 := \mathrm{span}\{v_1^{(v_0)},\ldots, v_{n_0}^{(v_0)}\}$$

(which is here identical with the geometric eigenspace). Then, $M_0 = P_{v_0} X$ with dim $M_0 = N_0$. Further, P_{V_0} can be calculated by $P_{V_0} = V_0 V_0^*$.

(cf. [11, pp. 234–238]). Let

$$\tilde{T}^{(1)} := P_{v_0} T^{(1)} P_{v_0}$$ (34)

and

$$v_k^{(1)}[\tilde{T}^{(1)}], \quad k = 1,\ldots,n_0$$

be the eigenvalues of $\tilde{T}^{(1)}$. Then,

$$v_1 := \max_{k=1,\dots,n_0} \{v_k^{(1)}[\tilde{T}^{(1)}] \mid \text{the associated eigenvector lies in } M_0\}. \tag{35}$$

Let

$$V_1 := [v_1^{(v_1)}, \dots, v_{n_1}^{(v_1)}]$$

be the matrix formed by the orthonormal set of eigenvectors $v_k^{(v_1)}, k = 1, \dots, n_1$ associated with v_1, and let Pv_1 be the orthogonal projection on the algebraic eigenspace

$$M_1 := \text{span}\{v_1^{(v_1)}, \dots, v_{n_1}^{(v_1)}\}.$$

Then, $M_1 = P_{v_1} X$ with $\dim M_1 = n_1$. As above, P_{V_1} can be calculated by

$$P_{v_1} = V_1 V_1^*$$

Let

$$\hat{T}^{(2)} := P_{v_1} \tilde{T}^{(2)} P_{v_1} := P_{v_1}(T^{(2)} - T^{(1)} S_{v_0} T^{(1)}) P_{v_1} \tag{36}$$

with

$$S_{v_0} := \sum_{v_k^{(0)} \neq v_0} \frac{1}{v_k^{(0)} - v_0} P_{v_k^{(0)}} \tag{37}$$

(for S_{V_0} cf. [6, p. 40, Problem 5.10, Formula (5.32)] and for $\tilde{T}^{(2)}$ cf. [6, p. 116]). Let

$$v_k^{(2)}[\hat{T}^{(2)}], \quad k = 1, \dots, n_1$$

be the eigenvalues of $\hat{T}^{(2)}$. Then,

$$v_2 := \max_{k=1,\dots,n_1} \{v_k^{(2)}[\hat{T}^{(2)}] \mid \text{the associated eigenvector lies in } M_1\} \tag{38}$$

Remark: In the formula for v_1, exactly those eigenvectors $v^{(1)}$ belong to M_1 for which $\operatorname{rank} P_{V_1} = \operatorname{rank}[P_{V_1}, v^{(1)}]$. Similarly, one proceeds in the formula for v_2.

From (32), we obtain the following theorem.

Theorem 4: Let $\Phi(t_0) \neq 0$. Then,

$$\|\Phi(t_0)\|_2 = v_0^{1/2} \tag{39}$$

$$D_+^1 \|\Phi(t_0)\|_2 = \tfrac{1}{2} v_1 / v_0^{1/2} \tag{40}$$

$$D_+^2 \|\Phi(t_0)\|_2 = \frac{1}{2} \frac{2 v_0 v_2 - \tfrac{1}{2} v_1^2}{v_0^{3/2}} \tag{41}$$

Proof: The proof follows from (1) and (32).

Case p=2: Second logarithmic derivative in the spectral norm

One can get $\mu_2^{(2)}[A]$ from the general result by setting $t_0 = 0$, i.e.,

$$\mu_2^{(2)}[A] = D_+^2 \|\Phi(0)\|_2 \tag{42}$$

in Theorem 4. But, from this formula one does not see any property. Therefore, we choose a different way and shall see that the formula gets very simple.

Let $\lambda(\hat{\Psi}(z))$ be an eigenvalue of $\hat{\Psi}(z)$ in (2). The aim is to find a local power series expansion of $\lambda(\hat{\Psi}(z))$ about $z_0 = 0$ up to the second order. From such a series, the coefficients of z (resp. $z^2/2$) deliver the first (resp. second) derivative of the function $z \mapsto \lambda(\hat{\Psi}(z))$ at $z_0 = 0$.

To achieve this, the first idea is to apply [6, Chapter II, Theorem 5.11, pp. 115–116, Formula (5.20)], where D_0 is chosen to be the whole z-plane.

A simpler way is to first rewrite $\lambda(\hat{\Psi}(z))$ in the form

$$\lambda(\hat{\Psi}(z)) = \lambda\left(E + B_1 z + \frac{B_2}{2}z^2 + \cdots\right)$$

$$= 1 + z\lambda\left(B_1 + z\frac{B_2}{2} + \cdots\right)$$

$$= 1 + z\lambda(T(z)) \tag{43}$$

with

$$T(z) = T + zT^{(1)} + \cdots \tag{44}$$

and

$$T = B_1,$$

$$T^{(1)} = \tfrac{1}{2}B_2. \tag{45}$$

This leads to the following lemma.

Lemma 5: There exists a number $t_0 > 0$ such that

$$\max_{j=1,\dots,n} \lambda_j(\Psi(t)) = \kappa_0 + \kappa_1 t + \kappa_2\frac{t^2}{2} + o(t^2), \quad 0 \leqslant t \leqslant t_0, \tag{46}$$

with

$$\kappa_0 := 1,$$

$$\kappa_1 := \lambda_{max} := \lambda_{max}(B_1), \tag{47}$$

$$\kappa_2 := \lambda_{max}^2$$

Proof: We want to apply [6, Chapter II, Theorem 5.4, Formula (5.10), p. 111]. For this, one has to show that the conditions of that theorem are fulfilled. Now,

(i)	$z \mapsto T(z)$ is differentiable at $z_0 = 0$, and
(ii)	each eigenvalue $\lambda(T)$ of T in (45) is semisimple because $T = B_1 = A^* + A$ is Hermitian.

Let P be the eigenprojection of T for λ (T),M=PX, and m=dim M. Further, let λ_i (T(z)) be the repeated eigenvalues of the λ(T)-group and $v_i^{(1)}$ the repeated eigenvalues of $PT'(0)P = P(\frac{1}{2}B_2)P$ in the subspace M for i=1,...,m. Moreover, let Cδ(0) be a circle about z_0=0 with sufficiently small radiusδ>0. Then, from [6, Chapter II, Theorem 5.4, Formula (5.10)] it follows that the λ (T)-group eigenvalues of T(z) have the form

$$\lambda_i(T(z)) = \lambda(T) + zv_i^{(1)} + o(z), \quad z \in C_\delta(0) \tag{48}$$

i=1,...,m. Setting $z = t \in C_\delta(0) \cap \mathbb{R}_0^+$, we obtain

$$\lambda_i(T(t)) = \lambda(T) + t\, v_i^{(1)} + o(t) \tag{49}$$

i=1,...,m. Let

$$\lambda(T) := \lambda_{\max}(T) := \max_{j=1,...,n} \lambda_j(T) \tag{50}$$

as well as

$$\lambda_{\max}(T(t)) := \max_{j=1,...,n} \lambda_j(T(t)) \tag{51}$$

Then, by a repeated application of [7, Lemma 2.1] we infer that a number t_0>0 with 0<t_0⩽δ exists such that

$$\lambda_{\max}(T(t)) = \lambda_{\max}(T) + t \max_{i=1,...,m} v_i^{(1)}(P\tfrac{1}{2}B_2P) + o(t), \quad 0 \leqslant t \leqslant t_0 \tag{52}$$

Since B_1 is Hermitian, so is the eigenprojection P because of [6, Chapter I, Section 5.3, Formula (5.22)]. Thus, P is orthogonal according to [6, Chapter I, Section 6.7]. Therefore, $\|x\|_2=\|Px\|_2=\|w\|_2$ forw:=Px, x∈X. So, due to [4, Section 71, p. 522], we have

$$\max_{i=1,...,m} v_i^{(1)} = \max_{i=1,...,m} v_i^{(1)}(P\tfrac{1}{2}B_2P) = \frac{1}{2} \sup_{\substack{x \in X \\ \|x\|_2=1}} (B_2Px, Px) = \frac{1}{2} \sup_{\substack{w \in M \\ \|w\|_2=1}} (B_2w, w) \tag{53}$$

Now, let $w \in M = PX, \|w\|_2 = 1$. Then, numbers $\alpha_k \in \mathbb{C}, k = 1,...,m$, exist such that $w = \sum_{k=1}^{m} \alpha_k v^{(k)}$ where $v^{(k)}$, k=1,...,m, are the linearly independent eigenvectors of $T=B_1$ corresponding to $\lambda_{max}(T)$. Set

$$\lambda_1 := \lambda_{max}(T) = \lambda_{max}(B_1)$$

(54)

and let $K_1 := \{1,....,m\}$. Then, because of $B_2 = A^*B_1 + B_1A$,

$$(B_2w, w) = \sum_{k \in K_1} \alpha_k(B_1v^{(k)}, Aw) + \sum_{k \in K_1} \overline{\alpha_k}(Aw, B_1v^{(k)})$$

$$= \sum_{k \in K_1} \alpha_k(\lambda_{max}v^{(k)}, Aw) + \sum_{k \in K_1} \overline{\alpha_k}(Aw, \lambda_{max}v^{(k)})$$

$$= \lambda_{max}\left\{\sum_{k \in K_1} \alpha_k(v^{(k)}, Aw) + \sum_{k \in K_1} \overline{\alpha_k}(Aw, v^{(k)})\right\}$$

$$= \lambda_{max}\{(A^*w, w) + (Aw, w)\}$$

$$= \lambda_{max}\{(B_1w, w)\} = \lambda_{max}\lambda_{max}$$

$$= \lambda_{max}^2$$

(55)

since λ_{max} is real. Thus,

$$\lambda_{max}(T(t)) = \lambda_{max}(B_1) + \frac{t}{2}\lambda_{max}^2(B_1) + o(t), \quad 0 \leqslant t \leqslant t_0$$

(56)

Consequently,

$$\lambda_{max}\left(E + B_1t + B_2\frac{t^2}{2} + \cdots\right)$$

$$= 1 + t\lambda_{max}(T(t))$$

$$= 1 + \lambda_{max}(B_1)t + \lambda_{max}^2(B_1)\frac{t^2}{2} + o(t^2)$$

(47)

which proves the lemma.

Now, we are in a position to determine the second logarithmic derivative of a real or complex n×n matrix A.

Theorem 6: Let $A \in \mathbb{R}^{n\times n}$ or $A \in \mathbb{C}^{n\times n}$. Then, the following formulae hold:

$$\mu_2^{(1)}[A] = \frac{1}{2}\kappa_1 = \lambda_{\max}\left(\frac{A^* + A}{2}\right) \tag{58}$$

and

$$\mu_2^{(2)}[A] = \frac{1}{2}\kappa_2 - \frac{1}{4}\kappa_1^2 = \lambda_{\max}^2\left(\frac{A^* + A}{2}\right) \tag{59}$$

In particular, the second logarithmic derivative in the spectral norm is always nonnegative.

Proof: The proof follows from the representation

$$\|\Phi(t)\|_2 = \left[1 + \kappa_1 t + \kappa_2 \frac{t^2}{2} + o(t^2)\right]^{1/2}, \quad 0 \leqslant t \leqslant t_0 \tag{60}$$

according to Lemma 5, and $\lambda_{\max} = \lambda_{\max}(B_1) = \lambda_{\max}(A^* + A)$. $\square$

Remark: The crucial point here is to have the idea that $\mu_2^{(2)}[A]$ might be simplified in such a dramatic way, and not to be content with Formula (42).

For normal matrices A, a general formula for the kth logarithmic derivative can be obtained.

Lemma 7: Let $A \in \mathbb{C}^{n\times n}$ be normal. Then, the kth logarithmic derivative in the spectral norm is given by

$$\mu_2^{(k)}[A] = \left[\lambda_{\max}\left(\frac{A^* + A}{2}\right)\right]^k = (\mu_2^{(1)}[A])^k, \quad k = 0, 1, 2, \ldots. \tag{61}$$

Proof: Since $A^*A = AA^*$, we have

$$B_j = (A^* + A)^j, \quad j = 0, 1, 2, \ldots \tag{62}$$

with $B_0 = E$. Thus,

$$\Psi(t) = \Phi^*(t)\,\Phi(t) = \sum_{j=0}^{\infty} \frac{(A^* + A)^j}{j!} t^j = e^{(A^* + A)t} \tag{63}$$

As $(A^* + A)$ is normal, according to [10, Eq. (4.78), p. 81]

$$\Psi(t) = V \operatorname{diag}\{e^{\lambda_j(A^* + A)t}\} V^{-1} \tag{64}$$

where $V = [v_1, \ldots, v_n]$ is the modal matrix whose columns consist of the eigenvectors of $B_1 = A^* + A$ pertinent to the eigenvalues of B_1, which is regular. Thus,

$$\lambda_j[\Psi(t)] = e^{\lambda_j(A^* + A)t}, \quad j = 1, \ldots, n \tag{65}$$

so that

$$\|\Phi(t)\|_2 = \left[\max_{j=1,\ldots,n} \lambda_j(\Psi(t)) \right]^{1/2} = \left[e^{\max_{j=1,\ldots,n} \lambda_j(A^* + A)t} \right]^{1/2} = e^{\lambda_{\max}((A^* + A)/2)t}, \quad t \geq 0 \tag{66}$$

Therefore, the assertion follows.

Remark: For normal matrices, even

$$D^k \|\Phi(t)\|_2 = \left[\lambda_{\max}\left(\frac{A^* + A}{2} \right) \right]^k \cdot e^{\lambda_{\max}((A^* + A)/2)t}, \quad t \geq 0, \quad k = 0, 1, 2, \ldots. \tag{67}$$

APPLICATIONS

In this section, we apply the obtained results

- to a vibratio n problem and get the best upper bounds in certain classes of upper bounds and

- to the discussion of the function $t \mapsto \|\Phi(t)\|p$, where $p \in \{\infty, 2\}$.

Upper Bounds for a Multi-mass Vibration Problem

1. The system matrix of the model. Consider the multi-mass vibration model in Fig. 1. The associated initial-value problem is given by

$$M\ddot{y} + B\dot{y} + Ky = 0, \quad y(0) = y_0, \quad \dot{y}(0) = \dot{y}_0$$

where $y = [y_1, \ldots, y_n]^T$ and

$$M = \begin{bmatrix} m_1 & & & & \\ & m_2 & & & \\ & & m_3 & & \\ & & & \ddots & \\ & & & & m_n \end{bmatrix}$$

$$B = \begin{bmatrix} b_1 + b_2 & -b_2 & & & & & \\ -b_2 & b_2 + b_3 & -b_3 & & & & \\ & -b_3 & b_3 + b_4 & -b_4 & & & \\ & & \ddots & \ddots & \ddots & & \\ & & & -b_{n-1} & b_{n-1} + b_n & -b_n & \\ & & & & -b_n & b_n + b_{n+1} \end{bmatrix}$$

$$K = \begin{bmatrix} k_1 + k_2 & -k_2 & & & & & \\ -k_2 & k_2 + k_3 & -k_3 & & & & \\ & -k_3 & k_3 + k_4 & -k_4 & & & \\ & & \ddots & \ddots & \ddots & & \\ & & & -k_{n-1} & k_{n-1} + k_n & -k_n & \\ & & & & -k_n & k_n + k_{n+1} \end{bmatrix}$$

or, in the state-space description

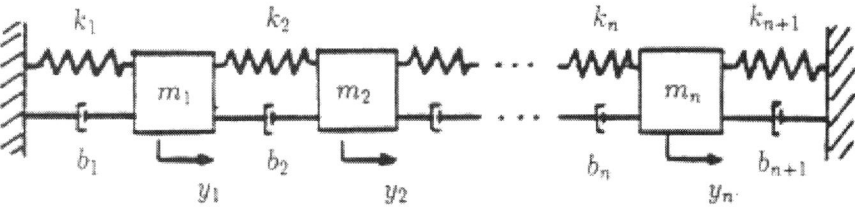

Figure 1: Multi-mass vibration model.

$$\dot{x}(t) = Ax(t), \quad x(0) = x_0$$

where $x = [y^T, z^T]^T, z = \dot{y}$, and where the system matrix A has the form

$$A = \left[\begin{array}{c|c} 0 & E \\ \hline -M^{-1}K & -M^{-1}B \end{array} \right]$$

As of now, we specify the values as

$$m_j = 1, \quad j = 1, \ldots, n,$$
$$k_j = 1, \quad j = 1, \ldots, n,$$

and

$$b_j = \begin{cases} 1/2, & j \text{ even,} \\ 1/4, & j \text{ odd.} \end{cases}$$

Then

M=E,

$$B = \begin{bmatrix} \frac{3}{4} & -\frac{1}{2} & & & & & \\ -\frac{1}{2} & \frac{3}{4} & -\frac{1}{4} & & & & \\ & -\frac{1}{4} & \frac{3}{4} & -\frac{1}{2} & & & \\ & & \ddots & \ddots & \ddots & & \\ & & & -\frac{1}{4} & \frac{3}{4} & -\frac{1}{2} \\ & & & & -\frac{1}{2} & \frac{3}{4} \end{bmatrix}$$

(if n is even), and

$$K = \begin{bmatrix} 2 & -1 & & & & \\ -1 & 2 & -1 & & & \\ & -1 & 2 & -1 & & \\ & & \ddots & \ddots & \ddots & \\ & & & -1 & 2 & -1 \\ & & & & -1 & 2 \end{bmatrix}$$

We choose n=5 in this paper.

2. The logarithmic derivatives. We obtain

$$\mu_\infty^{(1)}[A] = 4,$$

$$\mu_\infty^{(2)}[A] = -4.5,$$

$$\mu_2^{(1)}[A] \doteq 0.83751916260103,$$

$$\mu_2^{(2)}[A] \doteq 0.70143834772393,$$

whereby in passing the relationis $\mu_2^{(2)}[A] = (\mu_2^{(1)}[A])^2$ confirmed numerically.

3. The functions $\|\Phi(\cdot)\|_p, D^1 + \|\Phi(\cdot)\|_p, D^2 + \|\Phi(\cdot)\|_p$ and best upper bounds on .

The functions $\|\Phi(\cdot)\|_p$, $D_+^1\|\Phi(\cdot)\|_p$, and $D_+^2\|\Phi(\cdot)\|_p$, $p \in \{\infty, 2\}$, are shown in the Fig. 2, Fig. 3, Fig. 4, Fig. 5, Fig. 6 and Fig. 7, respectively. In Fig. 2 and Fig. 5, also upper bounds on $\|\Phi(\cdot)\|_\infty$ and $\|\Phi(\cdot)\|_2$ are plotted, respectively. The stepsize in all plots is Δt=0.1.

Best Upper Bound on $\|\Phi(\cdot)\|_\infty$ "Near the Origin"
Since $\mu_\infty^{(2)}[A] < 0$, it is clear that $\|\Phi(t)\|_\infty$ is bounded from above by the tangent at t=0; more precisely, there exists a number $t_0 > 0$ such that

$$\|\Phi(t)\|_\infty \leq 1 + \mu_\infty^{(1)}[A]\, t, \quad 0 \leq t \leq t_0$$

From the Fig. 2, Fig. 3 and Fig. 4, it follows that this inequality even holds for all t≥0. Apparently, this upper bound is the best one in the

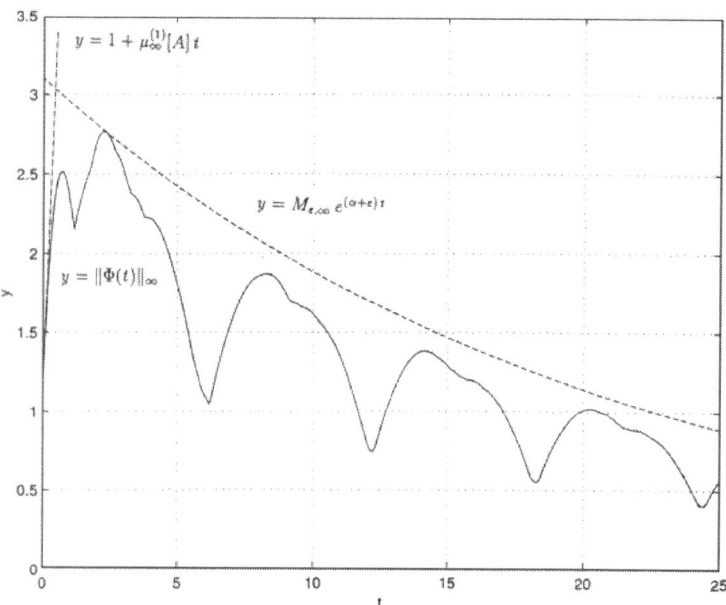

Figure 2: Chebyshev norm of the fundamental matrix and upper bounds.

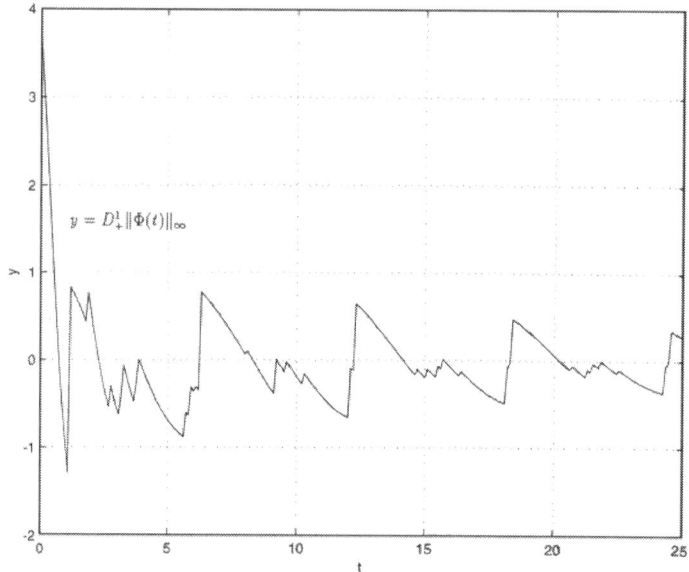

Figure 3: First right derivative of the Chebyshev norm of the fundamental matrix.

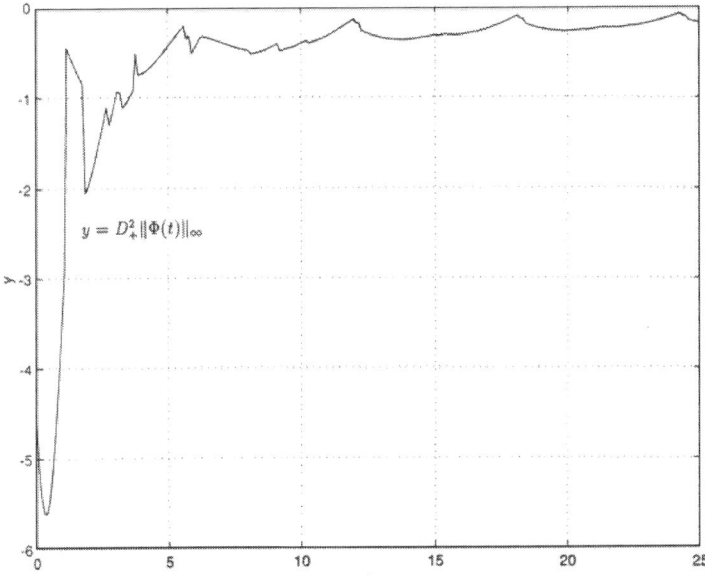

Figure 4: Second right derivative of the Chebyshev norm of the fundamental matrix.

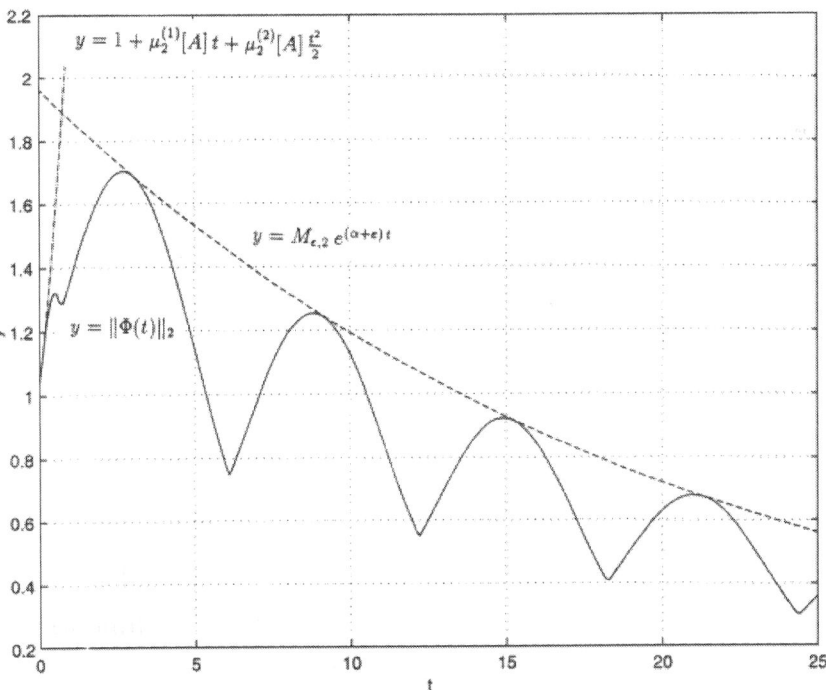

Figure 5: Spectral norm of the fundamental matrix and upper bounds.

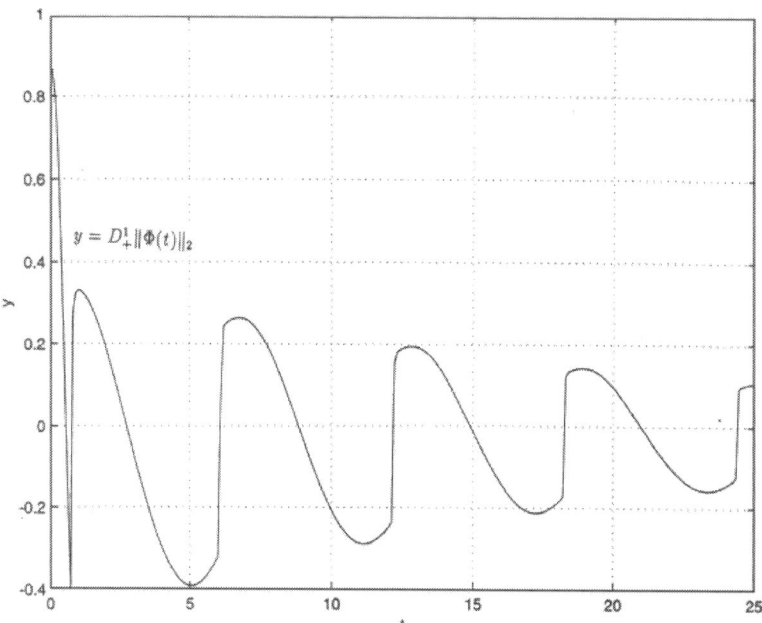

Figure 6: First right derivative of the spectral norm of the fundamental matrix.

class of polynomials $y = \sum_{k=1}^{N} \mu_\infty^{(k)}[A](t^k / k!)$ with $N \in \mathbb{N}$ and better than $e^{\mu_\infty^{(1)}[A]t}$.

Best Upper Bound on $\|\Phi(\cdot)\|_\infty$ "Far from the Origin"

As is well-known, for every $\varepsilon > 0$ there exists a constant such that

$$\|\Phi(t)\|_\infty \leqslant M_{\varepsilon,\infty} \, e^{(\alpha+\varepsilon)t}, \qquad t \geqslant 0$$

where α is the spectral abscissa of the matrix A. In this case, one could even choose $\varepsilon=0$ since the index of $\lambda_{max}(A)$ equals 1. But, in the programs we have choosen the machine precision ε=eps=$2^{-52} \doteq 2.2204 10^{-16}$ of MATLAB in order not to be bothered by this question.

Let $\|\cdot\|$ be any matrix operator norm. To obtain the minimal M_ε such that $\|\Phi(t)\| \leq M_\varepsilon e^{(\alpha+\varepsilon)t}, t \geq 0$ we seek a place t_c at which the function

$$t \mapsto \varphi_{M_\varepsilon}(t) := M_\varepsilon \, e^{(\alpha+\varepsilon)t}, \qquad t \geqslant 0$$

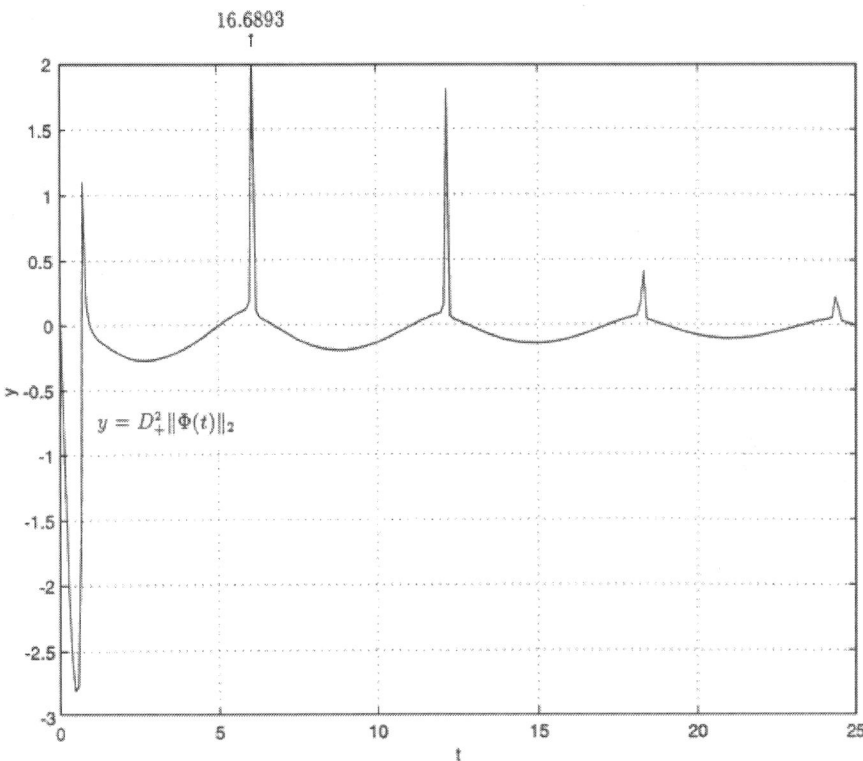

Figure 7: Second right derivative of the spectral norm of the fundamental matrix.

meets the function $t \mapsto \|\Phi(t)\|$. Thus,

$$\|\Phi(t_c)\| \overset{!}{=} \varphi_{M_\varepsilon}(t_c) \tag{68}$$

And

$$D_+^1 \|\Phi(t_c)\| \overset{!}{=} \varphi'_{M_\varepsilon}(t_c) = (\alpha + \varepsilon)\varphi_{M_\varepsilon}(t_c) \tag{69}$$

This is a system of two nonlinear equations in the two unknowns t_c and M_e. By eliminating $\Phi M_e(t_c)$, this system is reduced to

$$D_+^1 \|\Phi(t_c)\| \overset{!}{=} (\alpha + \varepsilon)\|\Phi(t_c)\| \tag{70}$$

which is a single nonlinear equation in the single unknown t_c. After t_c has been determined from (70), we get M_e in the following way: As

$$M_\varepsilon \, e^{(\alpha+\varepsilon)\,t_c} = \varphi_{M_\varepsilon}(t_c) = \|\Phi(t_c)\|$$

we have

$$M_\varepsilon = \|\Phi(t_c)\| \, e^{-(\alpha+\varepsilon)t_c}$$

This is now applied to the case p=∞. There are many points $t_{c,\infty}$, for which (70) is fulfilled. But only near $t_{c,\infty}$=2.5, i.e., near the largest peak, the calculated $M_{e,\infty}$ delivers an upper bound for t≥0. We obtain

$$\alpha \doteq -0.0502,$$

$$t_{c,\infty} \doteq 2.3930,$$

$$M_{\varepsilon,\infty} \doteq 3.1148.$$

It has been checked numerically down to t=700, where $\|\Phi(t)\|_\infty \approx 10^{-15}$, that the curve

$$y = M_{\varepsilon,\infty} \, e^{(\alpha+\varepsilon)\,t}$$

is actually an upper bound on $y=\|\Phi(t)\|_\infty$. The ordinary method to determine the constant $M_{e,\infty}$ starts from the representation

$$\Phi(t) = X \, e^{A\,t} X^{-1}$$

where X is the modal matrix, i.e., the matrix of eigenvectors and where e^{At} is the diagonal matrix $\mathrm{diag}(e^{\lambda_j t})$ with $\lambda_j, j=1,....,n$, being the eigenvalues of A, provided that A can be diagonalized (which is the case here). From this, we infer

$$M_{\varepsilon,\infty} = \|X\|_\infty \|X^{-1}\|_\infty \doteq 13.3640$$

which is worse than the constant $M_{e,\infty}$ determined by the differential calculus developed in this paper.

The intersection point t_∞^* between $y=1+\mu_\infty^1[A]t$ and $y=\varphi_{M_{\varepsilon,\infty}}(t)=M_{\varepsilon,\infty}e^{(\alpha+\varepsilon)t}$ is determined as $t_\infty^* = 0.50904644$

Best Upper Bound on $\|\Phi(\cdot)\|_2$ "Near the Origin"

Since $\mu_2^{(2)}[A]=(\mu_2^{(1)}[A])^2$, we always have that in the class of upper bounds on $\|\Phi(t)\|_2$ in the form of polynomials $y = \sum_{k=1}^{N}\mu_\infty^{(k)}[A](t^k/k!)$ with $N \in \mathbb{N}$, the tangent at $t=t_0=0$ is never possible; we need at least one term more. From Fig. 5, Fig. 6 and Fig. 7, we see that here we have

$$\|\Phi(t)\|_2 \leqslant 1 + \mu_2^{(1)}[A]t + \mu_2^{(2)}[A]\frac{t^2}{2}, \quad t \geqslant 0$$

Best Upper Bound on $\|\Phi(\cdot)\|_2$ "Far from the Origin"

By the differential calculus explained above, we get for the constant $M_{e,2}>0$ with

$$\|\Phi(t)\|_2 \leqslant M_{\varepsilon,2}\, e^{(\alpha+\varepsilon)t}, \quad t \geqslant 0$$

the values

$t_{c,2} \doteq 3.0507,$

$M_{\varepsilon,2} \doteq 1.9689.$

It has been checked numerically as well down to t=700, where $\|\Phi(t)\|_2 \approx 2\times10^{-15}$, that the curve

$$y = M_{\varepsilon,2}\, e^{(\alpha+\varepsilon)t}$$

is actually an upper bound on $y=\|\Phi(t)\|_2$. The ordinary method delivers

$$M_{\varepsilon,2} = \|X\|_2\, \|X^{-1}\|_2 \doteq 2.4860$$

which is worse than the constant $M_{e,2}$ determined by the differential calculus.

The intersection point t_2^* between $y=1+\mu_2^{(1)}[A]t+\mu_2^{(2)}[A](t^2/2)$ and $y=\varphi_{M_{\varepsilon,2}}(t)=M_{\varepsilon,2}e^{(\alpha+\varepsilon)t}$ is determined as $t_2^* = 0.79765557$

4. Remarks on the used software. The programs are written in MAT-LAB. Specifically, the following MATLAB functions have been used: expm, norm, rank, eig, fsolve, inv, plot, grid, size, length, zeros,

abs, find, fprintf, break, and control clauses such as for ... end, if ... else ... end, and the like.

5. Remarks on the formula for v_1. For p=2, we have tried what happens when v_1 in (35) is replaced by the formula

$$\tilde{v}_1 := \max_{k=1,\dots,n_0} \{v_k^{(1)}[\tilde{T}^{(1)}]\}$$

In this case, instead of $D_+^1 \|\Phi(t)\|_2$ in Fig. 6 we obtain

$$D_+^1 \|\Phi(t)\|_{2,\text{trunc}} := \begin{cases} D_+^1 \|\Phi(t)\|_2, & D_+^1 \|\Phi(t)\|_2 \geq 0, \\ 0, & D_+^1 \|\Phi(t)\|_2 < 0, \end{cases}$$

which is not illustrated by a figure for the sake of brevity.

6. Remark on the bounds in the range $[0, t_p^*]$

In the above examples, the advantage of the upper bounds $y = 1 + \mu_\infty^{(1)}[A]t$ (resp. $y = 1 + \mu_2^{(1)}[A]t + \mu_2^{(2)}[A](t^2/2)$ over $y = e^{\mu_\infty^{(1)}[A]t}$ (resp. $y = e^{\mu_2^{(1)}[A]t}$ in the interval $[0, t_\infty^*]$ (resp. $[0, t_2^*]$) is not great since t_∞^* (resp. t_2^*) are small. But, there are examples when this is different. For instance, if we choose n=25, then in the range $[0,25]$, the new best upper bounds lead to a great improvement over the old estimate $e^{\mu[A]t}$, which here is worthless for large t since it is by far too large. For the sake of brevity, we do not illustrate this by a plot.

On the other hand, t_p^* may be zero, in which case the new upper bounds on $[0, t_p^*]$ do not exist. For example, if n=1, A=[a], and B=[0], then y=eat is the solution to the problem $\dot{x} = ax, x(0) = 1$, for $t = t_2^* = t_\infty^* \geq$. In this case, $M_{e,\infty} = M_{e,2} = 1$.

Discussion of the Function $t \mapsto \|\Phi(t)\|_p$

The right derivatives obtained can also be used to make a discussion of the function $t \mapsto \|\Phi(t)\|_p, t \geq 0, p \in \{\infty, 2\}$. For example, the relative extrema and inflexion points of that function can be determined. This is important if the above mapping and its right derivatives are to be incorporated into a professional software.

As an example, in the case p=2, we obtain the following list of inflexion points in the range [0, 25] (with 4 digital places):

0.7117;1.0103;5.0487;6.7264;11.1410;12.8064;17.2267;18.8902;23.3105;24.9741.

In this example, it would be difficult to decide on grounds of Fig. 5 alone whether all these inflexion points exist.

CONCLUDING REMARKS

The differential calculus for p-norms developed in this paper has proved to be a powerful method to obtain better results in the field of upper bounds on the norm of the fundamental matrix.

However, further research is needed. For example, it would be of interest to prove strictly that one always obtains the best upper bound of the form $\|\Phi(t)\| \leq M_g e^{(\alpha+\varepsilon)t}, t \geq 0$, when choosing the point of contact t_c near the largest peak of $y=\|\Phi(t)\|$; here, we had to be content with a numerical check.

As an objection, one might say that the new method is limited to matrices of moderate size. In practice, however, in the case of large matrices A the engineers always use a process called reduction to obtain matrices of small size; for the Guyan reduction, the reader is referred to [9] and [5]. So, the new method can actually also be applied to very large problems.

REFERENCES

1. W.A. Coppel, Stability and Asymptotic Behavior of Di"erential Equations, D.C. Heath, Boston, 1965.
2. G. Dahlquist, Stability and Error Bounds in the Numerical Integration of Ordinary Di"erential Equations, Transactions of the Royal Institut of Technology, Stockholm, 1959.
3. Ch.A. Desoer, H. Haneda, The measure of a matrix as a tool to analyse computer algorithms for circuit analysis, IEEE Trans. Circuit Theory 19 (5) (1972) 480–486.
4. D.K. Faddejew, W.N. Faddejewa, Numerische Methoden der linearen Algebra, R. Oldenbourg Verlag, Munchen T Wien, 1976.
5. R.J. Guyan, Reduction of sti"ness and mass matrices, A.I.A.A. J. 3 (2) (1965) 380.
6. T. Kato, Perturbation Theory for Linear Operators, Springer, New York, 1966.
7. L. Kohaupt, Second logarithmic derivative of a complex matrix in the Chebyshev norm, SIAM J. on Matrix Anal. Appl. 21 (2) (1999) 382–389.
8. S.M. LozinskiWi, Error estimates for the numerical integration of ordinary di"erential equations (Russian), I. Izv. VysWs. UWcebn. Zaved. Matematika 6 (5) (1958) 52–90.
9. Msc=Nastran, The Theoretical Manual, pp. 3.5-6".
10. P.C. Muller, W.O. Schiehlen, Lineare Schwingungen, Akademische Verlagsgesellschaft, Wiesbaden, 1976. T
11. H. Niemeyer, E. Wermuth, Lineare Algebra, Vieweg, Braunschweig Wiesbaden, 1987.

CITATION

1. L. Kohaupt, Differential calculus for some p-norms of the fundamental matrix with applications, Journal of Computational and Applied Mathematics, Volume 135, Issue 1, 1 October 2001, Pages 1-21, ISSN 0377-0427, http://dx.doi.org/10.1016/S0377-0427(00)00559-8.

Fractional Versions of the Fundamental Theorem of Calculus

Eliana Contharteze Grigoletto and Edmundo Capelas de Oliveira
Department of Applied Mathematics,
University of Campinas, Campinas, Brazil

ABSTRACT

The concept of fractional integral in the Riemann-Liouville, Liouville, Weyl and Riesz sense is presented. Some properties involving the particular Riemann-Liouville integral are mentioned. By means of this concept we present the fractional derivatives, specifically, the Riemann-Liouville, Liouville, Caputo, Weyl and Riesz versions are discussed. The so-called fundamental theorem of fractional calculus is presented and discussed in all these different versions.

INTRODUCTION

Fractional calculus, a popular name used to denote the calculus of non-integer order, is as old as the calculus of integer order as created independently by Newton and Leibniz. In contrast with the calculus of integer order, fractional calculus has been granted a specific area of mathematics only in 1974, after the first international congress dedicated exclusively to it. Before this congress there were only sporadic independent papers, without a consolidated line [1, 2].

During the 1980s fractional calculus attracted researchers and explicit applications began to appear in several fields. We mention the doctoral thesis, published as an article [3], which seems to be the first one in the subject and the classical book by Miller and Ross [1], where one can see a timeline from 1645 to 1974. After the decade of 1990, completely consolidated, there appeared some specific journals and several textbooks were published. These facts lent a great visibility to the subject and it gained prestige around the world. An interesting timeline from 1645 to 2010 is presented in references [4-6]. We recall here that an important advantage of using fractional differential equations in applications is their non-local property. The use of fractional calculus is more realistic and this is one reason why fractional calculus has become more popular.

Nowadays, fractional calculus can be considered a frontier area in mathematics in the sense that there is as much research on its applications as there is on the calculus of integer order. Several applications in all areas of knowledgement are collected, presented and discussed in different books as follow [7-12].

The main objective of this paper is to explain what is meant by calculus of non-integer order and collect any different versions of the fractional derivatives associated with a particular fractional integral. Specifically, we recover the concepts of fractional integral and fractional derivative in different versions and present a new version of the so-called fundamental theorem of fractional calculus (FTFC), which is interpreted as a generalization of the classical fundamental theorem of calculus. We mention three recent works where FTFC is discussed, Tarasov's book [12], a paper by Tarasov [13] and a paper by Dannon [14] in which a particular case of the parameter associated with the derivative is presented. The paper is written as follows: in section two, we first review the concept of fractional integral in the Riemann-Liouville sense, which can be interpreted as a generalization of the integral of integer order and in the Liouville sense, which is a particular case of the Riemann-Liouville one. We review also the concept of fractional integral in the Weyl sense and in the Riesz sense. Section two present also the concepts of derivative as proposed by Riemann-Liouville, Liouville, Caputo, Weyl and Riesz, showing the real importance and applications. Some properties are also presented, among which one associated with the semigroup property. Our main result appears in section three, in which we present and demonstrate the many faces of the FTFC, in all different versions and which are interpreted as a generalization of the fundamental theorem of calculus. Applications are presented in section four.

FRACTIONAL CALCULUS

The integral and derivatives of non-integer order have several applications and are used to solve problems in different fields of knowledge, specifically, involving a fractional differential equation with boundary value conditions and/or initial conditions [7,9,11,12]. They can be seen as generalizations of the integral and derivatives of integer order. On the other hand, we mention two papers, by Heymans & Podlubny [15] and Podlubny [16] that provide an interesting geometric interpretation, and discuss applications of fractional calculus, with integral and derivatives of non-integer order. Also, we mention a recent paper in which the authors discuss a fractional differential equation with integral boundary value conditions [17]. We remember that, there are several ways to introduce the concepts of fractional integral and fractional derivatives, which are not necessarily coincident with each other [18]. The so-called Grünwald-Letnikov derivative, which will be not discussed in this paper, is convenient and useful to affront problems involving a numerical treatment [19].

In this section, the concept of fractional integral in the Riemann-Liouville, Liouville, Weyl and Riesz sense is presented. Some properties involving the particular Riemann-Liouville integral are mentioned. By means of this concept we present the fractional derivatives, specifically, the Riemann-Liouville, Liouville, Caputo, Weyl and Riesz versions are discussed.

Fractional Integral of Riemann-Liouville

The fractional integral of Riemann-Liouville is an integral that generalizes the concept of integral in the classical sense, and which can be obtained as a generalization of the Cauchy-Riemann integral. As we have already said, before we define the fractional integral Riemann-Liouville.

Definition 1: (Spaces $I_{a+}^{\alpha}\left(L_p\right)$ and $I_{a-}^{\alpha}\left(L_p\right)$) The spaces $I_{a+}^{\alpha}\left(L_p\right)$ and $I_{a-}^{\alpha}\left(L_p\right)$ are defined, for Re $(\alpha)>0$ and $p \geq 1$, by

$$I_{a+}^{\alpha}\left(L_p\right):=\left\{f(x): f(x)=I_{a+}^{\alpha}g(x), g(x) \in L^p(a,b)\right\}$$

and

$$I_{a-}^{\alpha}\left(L_p\right):=\left\{f(x): f(x)=I_{a-}^{\alpha}h(x), h(x) \in L^p(a,b)\right\}$$

respectively.

Property 2.1: (Semigroup) If Re $(\alpha)>0$ or $\alpha=0$ and if Re $(\beta)>0$ or $\beta=0$, then $I_{a+}^{\alpha}\ I_{a+}^{\beta}=I_{a+}^{\alpha+\beta}$ and $I_{b-}^{\alpha}\ I_{b-}^{\beta}=I_{b-}^{\alpha+\beta}$.

(Riemann-Liouville integrals) Let1 $f(x) \in L^1(a,b)$ with $-\infty<a<b<\infty$. The fractional integrals of Riemann-Liouville of order $\alpha \in \mathbb{C}$ with Re $(\alpha)>0$, on the left and on the right, are defined respectively by

$$\left(I_{a+}^{\alpha}f\right)(x):=\frac{1}{\Gamma(\alpha)}\int_a^x \frac{f(\tau)}{(x-\tau)^{1-\alpha}}d\tau, \text{with} x > a,$$

and

$$\left(I_{b-}^{\alpha}f\right)(x):=\frac{1}{\Gamma(\alpha)}\int_x^b \frac{f(\tau)}{(\tau-x)^{1-\alpha}}d\tau, \text{with} x < b.$$

If $\alpha=0$, we have $I_a^0 \equiv I_{b-}^0 := I$, where I is the identity operator.

Fractional Derivative of Riemann-Liouville

After we introduce the fractional integral in the Riemann-Liouville sense, we define the fractional derivative of Riemann-Liouville, which is the most used by mathematicians, particularly, into the problems in which the initial conditions are not involved.

(Riemann-Liouville derivative) Let $\alpha \in \mathbb{C}$, with Re $(\alpha)>0$ and n=[Re $(\alpha)]+1$, where $[\mu]$denotes the integer part of μ, the fractional derivatives in the Riemann-Liouville sense, on the left and on the right, are defined by

$$\left(D_{a+}^{\alpha}f\right)(x):=\frac{d^{n}}{dx^{n}}\left[\left(I_{a+}^{n-\alpha}f\right)(x)\right]$$

(1)

and

$$\left(D_{b-}^{\alpha}f\right)(x):=(-1)^{n}\frac{d^{n}}{dx^{n}}\left[\left(I_{b-}^{n-\alpha}f\right)(x)\right],$$

(2)

respectively.

Note that the derivatives in Equations (1) and (2) exist for[2] $f(x) \in AC^{n}$ (a, b).

If, in particular, $\alpha = n \in \mathbb{N}^{*}$, then

$$\left(D_{a+}^{n}f\right)(x)=f^{(n)}(x)$$

and

$$\left(D_{b-}^{n}f\right)(x)=(-1)^{n}f^{(n)}(x).$$

Fractional Integral and Derivative in the Liouville Sense

An interesting particular case of the fractional integral of Riemann-Liouville and the corresponding fractional derivative, is the so-called Liouville fractional integral and the Liouville fractional derivative. This case is obtained by substitution $a \to -\infty$ and $b \to \infty$ in the expressions associated with the fractional integral in the Riemann-Liouville sense.

(Liouville integral and derivative) The fractional integrals in the Liouville sense on the real axis, on the left and on the right, for $x \in \mathbb{R}$ and Re $(\alpha)>0$, are defined by

$$\left(I_+^\alpha f\right)(x) = \frac{1}{\Gamma(\alpha)} \int_{-\infty}^{x} \frac{f(\tau)}{(x-\tau)^{1-\alpha}} d\tau$$

(3)

and

$$\left(I_-^\alpha f\right)(x) = \frac{1}{\Gamma(\alpha)} \int_{x}^{\infty} \frac{f(\tau)}{(\tau-x)^{1-\alpha}} d\tau,$$

(4)

respectively.

The corresponding fractional derivatives in the Liouville sense are given by

$$\left(D_+^\alpha f\right)(x) := \frac{d^n}{dx^n}\left[\left(I_+^{n-\alpha} f\right)(x)\right]$$

(5)

and

$$\left(D_-^\alpha f\right)(x) := (-1)^n \frac{d^n}{dx^n}\left[\left(I_-^{n-\alpha} f\right)(x)\right],$$

(6)

where N=[Re $(\alpha)>0$] + 1, Re $(\alpha)>0$ and x $\in \mathbb{R}$

Fractional Integral and Derivative in the Weyl Sense

The operations involving the fractional integral and the fractional derivative, as above defined by means of the Riemann-Liouville operators are convenient for a function represented by a series of power but not for functions defined, for example, by means of Fourier series, because f(x) is a periodic function with period 2π the $(I_{a+}^\alpha f)(x)$ cannot be periodic. For this reason, it is convenient to introduce the fractional integral and the fractional derivatives in the so-called Weyl sense. First, some remarks on the Fourier series are presented.

Let f(x) be a periodic function with period 2π, defined on the real axis, with null average value, i.e.,

$$\frac{1}{2\pi}\int_0^{2\pi} f(x) dx = 0,$$

and

$$f(x) \sim \sum_{n=-\infty}^{\infty} c_n e^{inx}, \text{ with } c_n = \frac{1}{2\pi} \int_0^{2\pi} e^{-inx} f(x) dx,$$

representing the Fourier series of f(x) where c_n are the corresponding Fourier coefficients. Note that, by hypotesis, as the function has null average value, we have $c_0 = 0$.

(Weyl integral and derivative) We define the fractional integral and the respective fractional derivatives in the Weyl sense by

$$^{W}I^{\alpha} f(x) \sim \sum_{\substack{n=-\infty \\ n \neq 0}}^{\infty} c_n (in)^{-\alpha} e^{inx}$$

and

$$W^{\alpha} f(x) \sim \sum_{\substack{n=-\infty \\ n \neq 0}}^{\infty} c_n (in)^{\alpha} e^{inx},$$

respectively, with $\alpha \in \mathbb{C}$ and Re $(\alpha) > 0$.

In the particular case $0 < \alpha < 1$, if $f(x) \in L^1 (0, 2\pi)$ then $^{W}I^{\alpha} f(x) = I_+^{\alpha} f(x)$ and $_{-\infty}W_x^{\alpha} f(x) = D_+^{\alpha} f(x)$ where I_+^{α} and D_+^{α} are defined in Equations (3) and (5), respectively. If we define the Fourier series of f(x) in the form

$$f(x) \sim \sum_{\substack{n=-\infty \\ n \neq 0}}^{\infty} c_n e^{-inx}, \text{ with } c_n = \frac{1}{2\pi} \int_0^{2\pi} e^{inx} f(x) dx,$$

then we obtain for $0 < \alpha < 1$, in the same way, the fractional integral in the Weyl sense on the right and the fractional derivative on the right, defined by

$$^{W}I^{\alpha} f(x) = I_-^{\alpha} f(x)$$

and

$$_{x}W_{\infty}^{\alpha} f(x) = D_-^{\alpha} f(x).$$

respectively.

Fractional Derivative in the Caputo Sense

The differential operator of non-integer order in the Caputo sense is similar to the differential operator of non-integer order in the Riemann-Liouville sense. The capital difference is that: in the Caputo sense, the derivative acts first on the function, after we evaluate the integral and in the Riemann-Liouville sense, the derivatives acts on the integral, i.e., we first evaluate the integral and after we calculate the derivative. The derivative in the Caputo sense is more restritive than the Riemann-Liouville one. We also note that, both derivatives are defined by means of the Riemann-Liouvile fractional integral. The importance associated with this derivative is that, the derivative in the Caputo sense can be used, for example, in the case of a fractional differential equation with initial conditions which have a well-known interpretation, as in the calculus of integer order [20, 21].

(Caputo derivative) Let $f(x) \in AC^n$ [a, b], with $-\infty < a < b < \infty$, $\alpha \in \mathbb{C}$ for Re $(\alpha) \geq 0$ and n= [Re (α)] + 1. The fractional derivatives on the left, $_c D_{a+}^\alpha$ and on the right, $_c D_{b-}^\alpha$, in the Caputo sense, are defined in terms of the fractional integral operator of Riemann-Liouville as

$$\left(_c D_{a+.}^\alpha f \right)(x) := \left(I_{a+}^{n-\alpha} f^{(n)} \right)(x)$$

(7)

and

$$\left(_c D_{b-}^\alpha f \right)(x) := (-1)^n \left(I_{b-}^{n-\alpha} f^{(n)} \right)(x).$$

For $\alpha = 0$ we have $_c D_{a+}^0 \equiv {}_c D_{b-}^0 := I$. In particular, if $a = n \in \mathbb{N}^*$, then we have

$$\left(_c D_{a+.}^n f \right)(x) = f^{(n)}(x)$$

and

$$\left(_c D_{b-}^n f \right)(x) = (-1)^n f^{(n)}(x).$$

Fractional Integral in the Riesz Sense

First of all, we introduce the fractional integral and the corresponding fractional derivative in the Euclidean space $\mathbb{R}^n$, but for our purpose we discuss specifically the case $f : \mathbb{R} \to \mathbb{R}$, only. The operations of fractional integral and fractional derivative in the Euclidean space $\mathbb{R}^n$ are fractional powers of the Laplacian operator, $(-\Delta)^{\alpha/2}$. For $\alpha \in \mathbb{C} \setminus \{0\}$ and functions $f(x)$, "sufficiently good" [9], with $x \in \mathbb{R}^n$, the fractional operator

$(-\Delta)^{\alpha/2} f(x)$ is defined in terms of the Fourier transform by $(-\Delta)^{\alpha/2} f = F^{-1} ||w||^{-\alpha} Ff = \mathbb{I}^\alpha f$, with $\text{Re}(\alpha) > 0$.

The so-called fractional integral of order α in the Riesz sense, denoted by $\mathbb{I}^\alpha$, which is also known by Riesz potential and is defined by the Fourier convolution product

$$\left(\mathbb{I}^\alpha f\right)(\mathbf{x}) = \int_{\mathbb{R}^n} K_\alpha(\mathbf{x} - \xi) f(\xi) \, d\xi,$$

with $\text{Re}(\alpha) > 0$, and

$$K_\alpha(\mathbf{x}) := \frac{1}{\gamma_n(\alpha)} \begin{cases} \|\mathbf{x}\|^{\alpha-n}, & \alpha - n \neq 0, 2, \cdots, \\ \|\mathbf{x}\|^{\alpha-n} \log\left(\frac{1}{\|\mathbf{x}\|}\right), & \alpha - n = 0, 2, \cdots, \end{cases}$$

is the Riesz kernel and $\gamma_n(\alpha)$ is defined in [9], as follows

$$\frac{\gamma_n(\alpha)}{2^\alpha \pi^{\frac{n}{2}} \Gamma(\alpha/2)} = \begin{cases} \left[\Gamma\left(\frac{n-\alpha}{2}\right)\right]^{-1}, & \text{with } \alpha - n \neq 0, 2, 4, \cdots, \\ (-1)^{\frac{n-\alpha}{2}} 2^{-1} \Gamma\left(1 + \frac{\alpha-n}{2}\right), & \text{with } \alpha - n = 0, 2, 4, \cdots \end{cases}$$

(Riesz integral) As we have already said, we take in particular, $f : \mathbb{R} \to \mathbb{R}$, then

$$\left(\mathbb{I}^{\alpha} f\right)(x) = \frac{\Gamma\left(\dfrac{1-\alpha}{2}\right)}{2^{\alpha}\pi^{\frac{1}{2}}\Gamma\left(\dfrac{\alpha}{2}\right)} \int_{-\infty}^{\infty} f(\xi)|x-\xi|^{\alpha-1}\,d\xi$$

(8)

$\alpha \neq 1, 3. 5, \cdots$ which is the Riesz fractional integral.

We can also write Equation (8) in terms of the Liouville integrals. For this end, we introduce a convenient convolution product, i.e., we re-write Equation (8) in the following form

$$\left(\mathbb{I}^{\alpha} f\right)(x) = c_{\alpha}\left(f * g\right)(x),$$

(9)

With

$$g(x) = |x|^{\alpha-1}, \quad c_{\alpha} = \frac{\Gamma\left(\dfrac{1-\alpha}{2}\right)}{2^{\alpha}\pi^{\frac{1}{2}}\Gamma\left(\dfrac{\alpha}{2}\right)}, \quad \alpha \neq 1,3,5,\cdots$$

Applying the Fourier transform in both sides of Equation (9), we get

$$\mathcal{F}\left[\mathbb{I}^{\alpha} f\right](\omega) = \mathcal{F}\left[c_{\alpha}\left(f * g\right)\right](\omega)$$

and

$$|\omega|^{-\alpha}\hat{f}(\omega) = c_{\alpha}\hat{f}(\omega)\hat{g}(\omega),$$

where the functions $\hat{f}(\omega)$ and $\hat{g}(\omega)$ are the Fourier transforms of the functions f(x) and g(x), respectively. Thus, we can write,

$$\hat{g}(\omega) = \frac{2^{\alpha} \pi^{\frac{1}{2}} \Gamma\left(\dfrac{\alpha}{2}\right)}{\Gamma\left(\dfrac{1-\alpha}{2}\right)} |\omega|^{-\alpha}.$$

Then, rewriting Equation (8) in terms of the Liouville integrals, we get

$$\mathbb{I}^{\alpha} f(x) = \frac{\Gamma\left(\dfrac{1-\alpha}{2}\right)}{2^{\alpha}\pi^{\frac{1}{2}}\Gamma\left(\dfrac{\alpha}{2}\right)} \int_{-\infty}^{\infty} f(\xi)|x-\xi|^{\alpha-1}\,d\xi = \frac{\Gamma\left(\dfrac{1-\alpha}{2}\right)}{2^{\alpha}\pi^{\frac{1}{2}}\Gamma\left(\dfrac{\alpha}{2}\right)} \left[\int_{-\infty}^{x} f(\xi)(x-\xi)^{\alpha-1}\,d\xi + \int_{x}^{\infty} f(\xi)(\xi-x)^{\alpha-1}\,d\xi \right]$$

$$= \frac{\Gamma(\alpha)\Gamma\left(\dfrac{1-\alpha}{2}\right)}{2^{\alpha}\pi^{\frac{1}{2}}\Gamma\left(\dfrac{\alpha}{2}\right)} \left[I_{+}^{\alpha} f(x) + I_{-}^{\alpha} f(x) \right] = \frac{1}{2\cos\left(\dfrac{\alpha\pi}{2}\right)} \left[I_{+}^{\alpha} f(x) + I_{-}^{\alpha} f(x) \right].$$

Finally, we can write the fractional integral in the Riesz sense, in terms of a sum of two Liouville integrals

$$\mathbb{I}^{\alpha} f(x) = \frac{1}{2\cos\left(\dfrac{\alpha\pi}{2}\right)} \left[I_{+}^{\alpha} f(x) + I_{-}^{\alpha} f(x) \right]$$

(10)

with $\alpha \neq 1, 3, 5, \cdots$ For the best of our knowledged this is a new result.

Fractional Derivative in the Riesz Sense

The fractional derivative in the Riesz sense has been introduced in problems that can be treat as a Fourier convolution product. In this section, we introduce this fractional derivative and express it in terms of the Liouville derivative. An example of a specific convolution product will be proved. As we have already mentioned, we present the general definition but we are interested in a particular case involving the parameter.

(Riesz derivative) The fractional derivative of f(x) in the Riesz sense, with $x \in \mathbb{R}^n$, is defined for Re $(\alpha) > 0$, by means of

$$\left(\mathbb{D}^{\alpha}f\right)(\mathbf{x}):=\frac{1}{d_{n}\left(l,\alpha\right)}\int_{\mathbb{R}^{n}}\frac{\left(\Delta_{\xi}^{l}f\right)(\mathbf{x})}{|\xi|^{n+\alpha}}d\xi \quad (l>\alpha),$$

(11)

Where d_{n} (l, α) and $(\Delta_{\xi}^{l}f)(x)$ are defined in [9]. The derivative in terms of the Fourier transform is

$$\left(-\Delta\right)^{\alpha/2}f=\mathcal{F}^{-1}\|\omega\|^{\alpha}\mathcal{F}f=\mathbb{D}^{\alpha}f.$$

In the particular case, f: $\mathbb{R}\rightarrow\mathbb{R}$, we have

$$\left(\mathbb{D}^{\alpha}f\right)(x)=k_{\alpha}\int_{-\infty}^{\infty}\frac{f(x)-f(x-\xi)}{|\xi|^{1+\alpha}}d\xi,$$

(12)

with 0< α <1 and

$$k_{\alpha}=\frac{2^{\alpha}\Gamma\left(1+\dfrac{\alpha}{2}\right)\Gamma\left(\dfrac{1+\alpha}{2}\right)\sin\left(\dfrac{\alpha\pi}{2}\right)}{\pi^{\frac{3}{2}}}.$$

Thus, considering 0< α <1, we can write Equation (12) in terms of the fractional derivative in the Liouville sense, as follows

$$\mathbb{D}^{\alpha}f(x)=\frac{1}{2\cos\left(\dfrac{\alpha\pi}{2}\right)}\left[D_{+}^{\alpha}f(x)+D_{-}^{\alpha}f(x)\right],$$

(13)

With 0< α <1.

In what follow we express the Riesz derivative $\mathbb{D}^{\alpha}$ f(x) in terms of a convolution product. Using Equation (13) we get

$$\mathbb{D}^{\alpha}f(x)=\frac{1}{2\cos\left(\dfrac{\alpha\pi}{2}\right)}\left[D_{+}^{\alpha}f(x)+D_{-}^{\alpha}f(x)\right]$$

$$= \frac{1}{2\cos\left(\frac{\alpha\pi}{2}\right)} \left[\frac{1}{\Gamma(1-\alpha)} \frac{d}{dx} \int_{-\infty}^{x} f(\xi)(x-\xi)^{-\alpha} d\xi + \frac{1}{\Gamma(1-\alpha)} \frac{d}{dx} \int_{x}^{\infty} f(\xi)(\xi-x)^{-\alpha} d\xi \right]$$

$$= \frac{1}{\Gamma(1-\alpha)2\cos\left(\frac{\alpha\pi}{2}\right)} \left[\frac{d}{dx} \int_{-\infty}^{x} f(\xi)(x-\xi)^{-\alpha} d\xi + (-1)^{-\alpha} \frac{d}{dx} \int_{x}^{\infty} f(\xi)(x-\xi)^{-\alpha} d\xi \right]$$

$$= \left[\frac{1+(-1)^{-\alpha}}{\Gamma(1-\alpha)2\cos\left(\frac{\alpha\pi}{2}\right)} \right] \frac{d}{dx} \int_{-\infty}^{x} f(\xi)(x-\xi)^{-\alpha} d\xi = \left[\frac{1+(-1)^{-\alpha}}{\Gamma(1-\alpha)2\cos\left(\frac{\alpha\pi}{2}\right)} \right] \int_{-\infty}^{x} f(\xi)\left[\frac{\partial}{\partial x}(x-\xi)^{-\alpha} \right] d\xi$$

$$= \left[\frac{1+(-1)^{-\alpha}}{\Gamma(1-\alpha)2\cos\left(\frac{\alpha\pi}{2}\right)} \right] \int_{-\infty}^{x} f(\xi)\left[-\alpha(x-\xi)^{-\alpha-1} \right] d\xi = \left[\frac{1+(-1)^{-\alpha}}{\Gamma(-\alpha)2\cos\left(\frac{\alpha\pi}{2}\right)} \right] \int_{-\infty}^{x} f(\xi)(x-\xi)^{-\alpha-1} d\xi.$$

Thus, the convenient convolution product is

$$\left(\mathbb{D}^{\alpha} f \right)(x) = d_{\alpha}\left(f * h \right)(x),$$

$$(14)$$

where h(x) = $x^{-\alpha-1}$, $d_{\alpha} = \left(\dfrac{1+(-1)^{-\alpha}}{\Gamma(-\alpha)2\cos\left(\dfrac{\alpha\pi}{2}\right)} \right)$ and $0< \alpha <1$.

Applying the Fourier transform in both sides of Equation (14), we obtain the Fourier transform of the function h(x) = $x^{-\alpha-1}$:

$$\mathcal{F}\left[\mathbb{D}^{\alpha} f \right](\omega) = \left(\frac{1+(-1)^{-\alpha}}{\Gamma(-\alpha)2\cos\left(\dfrac{\alpha\pi}{2}\right)} \right) \mathcal{F}\left[(f * h) \right](\omega)$$

$$|\omega|^{\alpha} \hat{f}(\omega) = \left(\frac{1+(-1)^{-\alpha}}{\Gamma(-\alpha)2\cos\left(\dfrac{\alpha\pi}{2}\right)} \right) \hat{f}(\omega)\hat{h}(\omega),$$

where $\hat{f}(\omega)$ and $\hat{h}(\omega)$ are the Fourier transforms of f(x) and h(x), respectively. Thus,

$$\hat{h}(\omega) = \left(\frac{\Gamma(-\alpha) 2\cos\left(\dfrac{\alpha\pi}{2}\right)}{1 + (-1)^{-\alpha}} \right) |\omega|^{\alpha},$$

(15)

with $0 < \alpha < 1$.

Using this result we prove a theorem involving the Fourier convolution of two particular functions.

Theorem 1 The Fourier convolution product of the functions $g(x) = |x|^{\alpha-1}$ and $h(x) = x^{-\alpha-1}$, with $0 < \alpha < 1$ is given by

$$(g*h)(x) = \frac{1}{c_\alpha d_\alpha} \delta(x),$$

(16)

With

$$c_\alpha = \frac{\Gamma\left(\dfrac{1-\alpha}{2}\right)}{2^\alpha \pi^{\frac{1}{2}} \Gamma\left(\dfrac{\alpha}{2}\right)}, \quad d_\alpha = \left(\frac{1 + (-1)^{-\alpha}}{\Gamma(-\alpha) 2\cos\left(\dfrac{\alpha\pi}{2}\right)} \right)$$

and

$$\frac{1}{c_\alpha d_\alpha} = \frac{4\Gamma(\alpha)\Gamma(-\alpha)\cos^2\left(\dfrac{\alpha\pi}{2}\right)}{1 + (-1)^{-\alpha}}$$

where $\delta(x)$ is the Dirac delta function.

Proof: Evaluating the Fourier transform of the function (g*h) (x), and using Equation (N) and Equation (15), we have

$$\mathcal{F}\left[(g*h)\right](\omega) = \hat{g}(\omega)\hat{h}(\omega)$$

$$= \left(\frac{|\omega|^{-\alpha}}{c_\alpha}\right)\left(\frac{|\omega|^\alpha}{d_\alpha}\right) = \frac{1}{c_\alpha d_\alpha} = \mathcal{F}\left[\frac{1}{c_\alpha d_\alpha}\delta\right](\omega).$$

Thus, the Fourier transform of the convolution product can be written as

$$\mathcal{F}\left[(g*h)\right](\omega) = \mathcal{F}\left[\frac{1}{c_\alpha d_\alpha}\delta\right](\omega).$$

To recover the convolution product, we apply the corresponding inverse Fourier transform in both sides of the last equation, and we get

$$(g*h)(x) = \frac{1}{c_\alpha d_\alpha}\delta(x)$$

$$= \left(\frac{4\Gamma(\alpha)\Gamma(-\alpha)\cos^2\left(\dfrac{\alpha\pi}{2}\right)}{1+(-1)^{-\alpha}}\right)\delta(x).$$

Note that, the coefficient $\dfrac{1}{c_\alpha d_\alpha}$ in Equation (16) is complex, because$0<$ $\alpha <1$.

THE FUNDAMENTAL THEOREM OF FRACTIONAL CALCULUS

After the presentation of different versions of the fractional integral operator and the corresponding fractional derivative it is natural to introduce the corresponding FTFC associated with these different versions. Then, we present in this section the so-called FTFC, in the Riemann-Liouville, Caputo, Liouville, Weyl and Riesz versions. The results that are known we mention the reference where one can see the proof, otherwise, we present the proof. As we have already said, in all cases we first write the theorem in general form, consider a particular case and finally, we recover, as a convenient limit, the fundamental theorem in the corresponding classical version.

Theorem 2(Riemann-Liouville): Consider a function f(x) such that f: [a, b] → $\mathbb{R}$, with $-\infty < a < b < \infty$; let $\alpha \in \mathbb{C}$ with Re $(\alpha) > 0$ and n=[Re (α)]+1. If f(x) ∈ AC^n [a, b] or C^n(a, b) then, for every x ∈ (a, b) we have:

1) $(D^{\alpha}_{a+}I^{\alpha}_{a+}f)(x) = f(x)$ and $(D^{\alpha}_{b-}I^{\alpha}_{b-}f)(x) = f(x)$.

2) For $\left(I^{n-\alpha}_{a+}f(x)\right) \in AC^n[a,b]$ we have

$$\left(I^{\alpha}_{a+}D^{\alpha}_{a+}f\right)(x)$$

$$= f(x) - \sum_{j=0}^{n-1} \frac{(x-a)^{\alpha-j-1}}{\Gamma(\alpha-j)}\left[\left(D^{n-j-1}I^{n-\alpha}_{a+}f(x)\right)\Big|_{x=a}\right],$$

$$(17)$$

and in the case $\left(I^{n-\alpha}_{b-}f(x)\right) \in AC^n[a,b]$, we have

$$\left(I^{\alpha}_{b-}D^{\alpha}_{b-}f\right)(x) = f(x)$$

$$-\sum_{j=0}^{n-1} \frac{(-1)^{n-j-1}(b-x)^{\alpha-j-1}}{\Gamma(\alpha-j)}\left[\left(D^{n-j-1}I^{n-\alpha}_{b-}f(x)\right)\Big|_{x=b}\right].$$

$$(18)$$

For f (x) ∈ I^{α}_{a+} (L^p) we have,

$$\left(I^{\alpha}_{a+}D^{\alpha}_{a+}f\right)(x) = f(x)$$

$$(19)$$

and for f (x)∈ I^{α}_{b-} (L$_p$) we have,

$$\left(I_{b-}^{\alpha} D_{b-}^{\alpha} f\right)(x) = f(x).$$

(20)

In particular, if $0 < \mathrm{Re}\,(\alpha) < 1$ in Equations (17) and (18), then

$$\left(I_{a+}^{\alpha} D_{a+}^{\alpha} f\right)(x) = f(x) - \frac{(x-a)^{\alpha-1}}{\Gamma(\alpha)}\left[\left(I_{a+}^{1-\alpha} f(x)\right)\Big|_{x=a}\right]$$

and

$$\left(I_{b-}^{\alpha} D_{b-}^{\alpha} f\right)(x) = f(x) - \frac{(b-x)^{\alpha-1}}{\Gamma(\alpha)}\left[\left(I_{b-}^{1-\alpha} f(x)\right)\Big|_{x=b}\right].$$

On the other hand, if $\alpha = 1$, we have

$$\left(I_{a+}^{1} D_{a+}^{1} f\right)(x) = f(x) - f(a)$$

and

$$\left(I_{b-}^{1} D_{b-}^{1} f\right)(x) = f(x) - f(b),$$

also[3]

$$\left({}_{a}I_{b}^{1} D_{a+}^{1} f\right)(x) = f(b) - f(a)$$

and

$$\left({}_{a}I_{b}^{1} D_{b-}^{1} f\right)(x) = f(a) - f(b).$$

Proof: (1) Both results follow from Lemma 2.4 in [9]. (2) To prove Equation (19) and Equation (20), we use Definition 1 and the case (1). If $f(x) \in I_{a+}^{\alpha}(L_p)$, then $f(x) \in I_{a+}^{\alpha} g(x)$. Thus, we can write,

$$\left(I_{a+}^{\alpha}D_{a+}^{\alpha}f\right)(x) = I_{a+}^{\alpha}\left[\left(D_{a+}^{\alpha}I_{a+}^{\alpha}g\right)(x)\right]$$

$$= \left(I_{a+}^{\alpha}g\right)(x) = f(x).$$

Now, if $f(x) \in I_{b-}^{\alpha}\left(L_{p}\right)$, it follows in an analogous way that

$$\left(I_{b-}^{\alpha}D_{b-}^{\alpha}f\right)(x) = f(x).$$

The Equations (17) and (18) follow from [11]. In the particular case $0 <$ Re $(\alpha) < 1$ we must substitute n=1 in Equations (17) and (18).

In the case $\alpha=1$ we recover

$$\left({}_{a}I_{b}^{1}D_{a+}^{1}f\right)(x) = \left[\left(I_{a+}^{1}D_{a+}^{1}f\right)(x)\right]_{x=b}$$

$$= \left[f(x) - f(a)\right]_{x=b} = f(b) - f(a)$$

and

$$\left({}_{a}I_{b}^{1}D_{b-}^{1}f\right)(x) = \left[\left(I_{b-}^{1}D_{b-}^{1}f\right)(x)\right]_{x=a}$$

$$= \left[f(x) - f(b)\right]_{x=a} = f(a) - f(b).$$

We will now show that the Theorem 2 in which we consider the fractional operator in the Caputo sense.

Theorem 3 (Caputo): Let f(x) be a function $f:[a, b] \to \mathbb{R}$, with $-\infty < a < b < \infty$ and let $\alpha \in \mathbb{C}$ with Re $(\alpha) > 0$ and n= [Re (α)] +1. If $f(x) \in AC^{n}$ [a, b] or C^{n} [a, b] then for $x \in$ (a, b)

1. For Re $(\alpha) \notin \mathbb{N}$ or $\alpha \in \mathbb{N}$, we have

$$\left({}_{c}D_{a+}^{\alpha}I_{a+}^{\alpha}f\right)(x) = f(x)$$

and

$$\left({}_C D_{b-}^\alpha I_{b-}^\alpha f\right)(x) = f(x).$$

2. We have

$$\left(I_{a+}^\alpha {}_C D_{a+}^\alpha f\right)(x) = f(x) - \sum_{j=0}^{n-1} \frac{(x-a)^j}{j!} \left(f^{(j)}(x)\Big|_{x=a}\right)$$

and

$$\left(I_{b-}^\alpha {}_C D_{b-}^\alpha f\right)(x)$$

$$= f(x) - \sum_{j=0}^{n-1} \frac{(-1)^j (b-x)^j}{j!} \left(f^{(j)}(x)\Big|_{x=b}\right).$$

In particular, if $0 < \mathrm{Re}\,(\alpha) < 1$, then

$$\begin{cases} \left(I_{a+}^\alpha {}_C D_{a+}^\alpha f\right)(x) = f(x) - f(a), \\ \left(I_{b-}^\alpha {}_C D_{b-}^\alpha f\right)(x) = f(x) - f(b), \end{cases}$$

and, if $\alpha = 1$, then

$$\left({}_a I_b^1 {}_C D_{a+}^1 f\right)(x) = f(b) - f(a)$$

and

$$\left({}_a I_b^1 {}_C D_{b-}^1 f\right)(x) = f(a) - f(b).$$

Proof: (1) It follows from Lemma 2.21 in [9]. (2) It follows from Lemma 2.22 in [9].

Theorem 4 (Liouville): Let f(x) be a function defined on real axis, $\alpha \in \mathbb{C}$ with Re(α)>0 and n = [Re(α)] +1. If f(x) $\in$ AC^n (-∞,∞) or C^n(-∞,∞) then, for x $\in \mathbb{R}$, we have:

1. $(D_+^\alpha I_+^\alpha f)(x) = f(x)$ and $(D_-^\alpha I_-^\alpha f)(x) = f(x)$

2) If 0< Re (α) <1 and $\lim\limits_{x \to -\infty} f(x) = 0 = \lim\limits_{x \to +\infty} f(x)$ then

$$\left(I_+^\alpha D_+^\alpha f\right)(x) = f(x)$$

and

$$\left(I_-^\alpha D_-^\alpha f\right)(x) = f(x).$$

Proof

1. Using Part (1) of the Theorem 2, follows

$$\left(D_+^\alpha I_+^\alpha f\right)(x) = \lim_{a \to -\infty} \left(D_{a+}^\alpha I_{a+}^\alpha f\right)(x) = \lim_{a \to -\infty} f(x) = f(x),$$

in the same way, we have $\left(D_-^\alpha I_{a-}^\alpha f\right)(x) = f(x)$.

2. Using Part (2) of the Theorem 2, we have that: if0< Re (α) <1, then

$$\left(I_+^\alpha D_+^\alpha f\right)(x)$$

$$= \lim_{a \to -\infty} \left\{ f(x) - \frac{(x-a)^{\alpha-1}}{\Gamma(\alpha)} \left[\left(I_{a+}^{1-\alpha} f(x)\right)\Big|_{x=a} \right] \right\} = f(x)$$

and

$$\left(I_-^\alpha D_-^\alpha f \right)(x)$$

$$= \lim_{b \to +\infty} \left\{ f(x) - \frac{(b-x)^{\alpha-1}}{\Gamma(\alpha)} \left[\left(I_{b-}^{1-\alpha} f(x) \right) \Big|_{x=b} \right] \right\} = f(x),$$

with the function f(x) → 0 when x → + ∞ or - ∞.

Theorem 5 (Weyl): Let f(x) ∈ $L^{1\,(0,\,2\pi)}$ be a periodic function with period 2π, defined on the real axis, with null average value, and let $\alpha \in \mathbb{C}$ with Re $(\alpha) > 0$ then, at $x \in \mathbb{R}$ in which the Fourier series of f(x) is convergent, we have

$$\left({}^W D^\alpha \, {}^W I^\alpha f \right)(x) = f(x) \text{ and } \left({}^W I^\alpha \, {}^W D^\alpha f \right)(x) = f(x).$$

Proof: Let

$$f(x) \sim \sum_{\substack{n=-\infty \\ n \neq 0}}^{\infty} c_n e^{inx}, \text{ with } c_n = \frac{1}{2\pi} \int_0^{2\pi} e^{-inx} f(x)\, dx,$$

be the Fourier series of(x), with the corresponding Fourier coefficients c_n, then

$$^W I^\alpha f(x) \sim \sum_{\substack{n=-\infty \\ n \neq 0}}^{\infty} c_n (in)^{-\alpha} e^{inx}$$

and

$$^W D^\alpha f(x) \sim \sum_{\substack{n=-\infty \\ n \neq 0}}^{\infty} c_n (in)^{\alpha} e^{inx},$$

with this, we have

$$\left({}^{W}I^{\alpha}\ {}^{W}D^{\alpha}f \right)(x) = {}^{W}I^{\alpha}\left[\sum_{\substack{n=-\infty \\ n\neq 0}}^{\infty} c_{n}\left(in \right)^{\alpha} e^{inx} \right]$$

$$= \sum_{\substack{n=-\infty \\ n\neq 0}}^{\infty} c_{n}\left(in \right)^{\alpha} \left(in \right)^{-\alpha} e^{inx} \sum_{\substack{n=-\infty \\ n\neq 0}}^{\infty} c_{n} e^{inx} \sim f(x).$$

In the same way, we have $\left({}^{W}D^{\alpha}\ {}^{W}I^{\alpha}\ f \right)(x) = f(x)$.

Theorem 6 (Riesz): Let $f(x) \in \Phi$, where Φ denote the so-called the space of Lizorkin functions, defined in [9], and let $\alpha > 0$, then

1) $\left(\mathbb{D}^{\alpha}\mathbb{I}^{\alpha}f \right)(x) = f(x).$

2) For $0 < \alpha < 1$ $(I^{\alpha}\ D^{\alpha}\ f) = f(x)$.

Proof: (1) See Property 2.35 in [9]. (2) For $0 < \alpha < 1$, using Equation (9) we have

$$\left(\mathbb{D}^{\alpha}\mathbb{I}^{\alpha}f \right)(x) = \mathbb{D}^{\alpha}\left[c_{\alpha}\left(f * g \right)(x) \right],$$

where

$$g(x) = \left| x \right|^{\alpha-1},$$

and

$$c_{\alpha} = \frac{\Gamma\left(\dfrac{1-\alpha}{2} \right)}{2^{\alpha}\,\pi^{\frac{1}{2}}\Gamma\left(\dfrac{\alpha}{2} \right)}.$$

Thus, using Equation (14), we get

$$\mathbb{D}^{\alpha}\left[c_{\alpha}\left(f*g\right)(x)\right]=c_{\alpha}d_{\alpha}\left[\left(f*g\right)*h\right](x),$$

with h(x) = x$^{-\alpha-1}$ and $d_{\alpha}=\left(\dfrac{1+(-1)^{-\alpha}}{\Gamma(-\alpha)\,2\cos\left(\dfrac{\alpha\pi}{2}\right)}\right)$

Using the Theorem 1, we obtain

$$\left(\mathbb{D}^{\alpha}\mathbb{I}^{\alpha}f\right)(x)=c_{\alpha}d_{\alpha}\left[\left(f*g\right)*h\right](x)$$
$$=c_{\alpha}d_{\alpha}\left[f*\left(g*h\right)\right](x)$$
$$=c_{\alpha}d_{\alpha}\left[f*\left(\frac{1}{c_{\alpha}d_{\alpha}}\delta\right)\right](x)$$

=f(x)

Considering $0<\alpha<1$, and using Equation (10) and Equation (13), we can write

$$\left(\mathbb{D}^{\alpha}\mathbb{I}^{\alpha}f\right)x=\frac{1}{2\cos\left(\dfrac{\alpha\pi}{2}\right)}\mathbb{D}^{\alpha}\left(I_{+}^{\alpha}f(x)+I_{-}^{\alpha}f(x)\right)$$

$$=\gamma^{2}\left[D_{+}^{\alpha}\left(I_{+}^{\alpha}f(x)+I_{-}^{\alpha}f(x)\right)+D_{-}^{\alpha}\left(I_{+}^{\alpha}f(x)+I_{-}^{\alpha}f(x)\right)\right]$$

$$=\gamma^{2}\left(D_{+}^{\alpha}I_{+}^{\alpha}f(x)+D_{+}^{\alpha}I_{-}^{\alpha}f(x)+D_{-}^{\alpha}I_{+}^{\alpha}f(x)+D_{-}^{\alpha}I_{-}^{\alpha}f(x)\right)$$

where $\gamma=\dfrac{1}{2\cos\left(\dfrac{\alpha\pi}{2}\right)}$ By means of the Theorem 4, for $0<\alpha<1$, we

can write $D_{+}^{\alpha}I_{+}^{\alpha}f(x)=D_{-}^{\alpha}I_{-}^{\alpha}f(x)=f(x)$. Thus, we can write

$$\left(\mathbb{D}^\alpha \mathbb{I}^\alpha f \right)(x)$$

$$= \gamma^2 \left[2f(x) + \mathrm{D}_+^\alpha \mathrm{I}_-^\alpha f(x) + \mathrm{D}_-^\alpha \mathrm{I}_+^\alpha f(x) \right].$$

On the other hand, using Theorem 6, for $0 < \alpha < 1$, we have $(\mathbb{D}^\alpha \ \mathbb{I}^\alpha f)$ $(x) = f(x)$. In this case, we can write,

$$f(x) = \frac{1}{4\cos^2\left(\dfrac{\alpha\pi}{2}\right)} \left(2f(x) + \mathrm{D}_+^\alpha \mathrm{I}_-^\alpha f(x) + \mathrm{D}_-^\alpha \mathrm{I}_+^\alpha f(x) \right),$$

or in the following form,

$$\mathrm{D}_+^\alpha \mathrm{I}_-^\alpha f(x) + \mathrm{D}_-^\alpha \mathrm{I}_+^\alpha f(x) = \left[2\cos^2\left(\frac{\alpha\pi}{2}\right) - 1 \right] f(x)$$

with $0 < \alpha < 1$.

Evaluating $(\mathbb{D}^\alpha \ \mathbb{I}^\alpha f)$, we get in the same way, that

$$\mathrm{I}_+^\alpha \mathrm{D}_-^\alpha f(x) + \mathrm{I}_-^\alpha \mathrm{D}_+^\alpha f(x) = \left[2\cos^2\left(\frac{\alpha\pi}{2}\right) - 1 \right] f(x)$$

with $0 < \alpha < 1$.

APPLICATIONS

In this section, using the FTFC, fractional differential equations are solved, one of them associated with the Riemann-Liouville case and the other involving the Caputo case.

Example 1: Consider the following fractional differential equation and its initial condition:

$$_C D_{0+}^{\alpha} y(t) = c, \text{ and } y(0) = 0,$$

with c a complex constant, t> 0 and $0 < \text{Re}(\alpha) < 1$.

Applying the fractional integral operator I_{0+}^{α} to the fractional differential equation and using Theorem 3, item (2), we can write

$$I_{0+}^{\alpha}\ _C D_{0+}^{\alpha} y(t) = I_{0+}^{\alpha} c \Leftrightarrow y(t) = \frac{ct^{\alpha}}{\Gamma(\alpha+1)}.$$

The next application, we discuss the same problem which has been discussed by Jafari & Momani [22] using another methodology, the so-called modified homotoy perturbation method. We solve the equation using the method of separation of variables and the FTFC (Riemann-Liouville).

Example 2: Consider the initial value problem involving the fractional diffusion equation

$$_C D_{0+}^{\alpha} u = -\Delta u, \text{ and } u(\overline{x}, 0) = e^{-(x_1 + x_2 + x_3)}, \tag{21}$$

Where $u \equiv u(\overline{x}, t), \overline{x} = (x_1, x_2, x_3)$ with $-\infty < x_i < \infty$ for i=1, 2, 3, t > 0 and a $(0, 1] \subset \mathbb{R}$.

Suppose a solution with the form

$$u(\overline{x}, t) = X(\overline{x}) T(t). \tag{22}$$

Substituting Equation (22) into the fractional diffusion equation, Equation (21), we get

$$\frac{_C D_{0+}^{\alpha} T(t)}{T(t)} = \frac{-\Delta X(\overline{x})}{X(\overline{x})} = \lambda,$$

where λ is a real constant.

We first consider the fractional differential equation $\dfrac{_c D_{0+}^{\alpha} T(t)}{T(t)} = \lambda$. Thus, we obtain

$$_c D_{0+}^{\alpha} T(t) = \lambda T(t).$$

$$(23)$$

Substituting Equation (7) into Equation (23), we get an equivalent equation

$$I_{0+}^{n-\alpha} T^{(n)}(t) = \lambda T(t).$$

Applying operator $_c D_{0+}^{n-\alpha}$ on both sides of the last equation we have

$$D_{0+}^{n-\alpha} I_{0+}^{n-\alpha} T^{(n)}(t) = D_{0+}^{n-\alpha}\left(\lambda T(t)\right).$$

Using Theorem 2, item (1), we can write

$$T^{(n)}(t) = \lambda D_{0+}^{n-\alpha} T(t).$$

$$(24)$$

As $\alpha \in (0, 1]$ we have n=1. We can also write $T^{(n)}(t) = _c D_{0+}^{n} T(t)$. Equation (24) can then be written

$$D_{0+}^{1} T(t) = \lambda D_{0+}^{1-\alpha} T(t).$$

$$(25)$$

This is a known equation and can be seen in reference [9], i.e., from Theorem 5.2, Equation (5.2.31) in [9] with $\alpha = 1$ and $\beta = 1-\alpha$, to obtain the result

$$T(t) = E_{\alpha}\left(\lambda t^{\alpha}\right),$$

$$(26)$$

Where $E_{\mu(x)}$ is the one-parameter Mittag-Leffler function.

Using the initial condition we have

$$X(\overline{x})T(0) = e^{-(x_1+x_2+x_3)},$$

and by Equation (26), T(0) =1 ,

then

$$X(\overline{x}) = e^{-(x_1+x_2+x_3)}.$$

Substituting this result in equation involving X ($\overline{x}$) we have – 3X ($\overline{x}$) = IX ($\overline{x}$), i.e., λ = -3.

Thus, the solution of the initial value problem, i.e., the fractional diffusion equation and the initial condition, is given by

$$u(\overline{x},t) = e^{-(x_1+x_2+x_3)}E_\alpha\left(-3t^\alpha\right).$$

(27)

We note that, in the paper by Jafari & Momani [22] its solution is presented with a misprint, i.e., as can be verified this solution is not a solution of Equation (21). We remark, in passing, that the solution presented in the paper by Jafari & Momani [22] is different from ours because it solution is not a solution of Equation (21).

As a particular case, we consider the problem associated with the unidimensional diffusion equation, that is

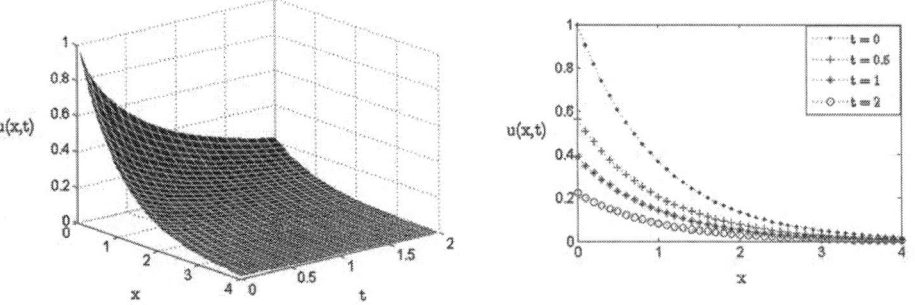

Figure 1:. Graphics of u(x, t) = e $^{-x}$ E$_\alpha$ (-t$^\alpha$) xx in the case α= 0.8.

$$_C D_{0+}^{\alpha} u = -\Delta u, \text{ and } u(x,0) = e^{-x},$$

Where $u \equiv u(x,t)$ with $-\infty < x < \infty$, $t>0$ and $\alpha \in (0, 1] \subset \mathbb{R}$.

In this case, the solution is $u(x, t) = e^{-x} E_{\alpha}(-t^{\alpha})$, since, $\lambda = -1$ in Equation (26). For $\alpha = 0.8$ the graphic is as in Figure 1.

CONCLUSIONS

After a brief introduction about the calculus of non-integer order, popularly known as fractional calculus, we presented the concept of fractional integral in the Riemann-Liouville sense. We then discussed the formulation of fractional derivatives as introduced by RiemannLiouville and, interchanging the integral with the derivative, we introduced the formulation proposed by Caputo. We presented also the fractional integral and fractional derivatives in the Liouville, Weyl and Riesz sense. As our main result, we colleted and showed the many faces of the FTFC, associated with the Riemann-Liouville, Caputo, Liouville, Weyl and Riesz version. As applications, we discussed two examples involving fractional differential equations. A natural continuation of this work resides in the fact that we can obtain solutions associated with fractional differential equations involving also fractional derivatives as proposed by Riesz and Weyl. A study in this direction is being developed [23].

ACKNOWLEDGMENTS

We are grateful to Dr. J. Emlio Maiorino and Dr. Quintino A. G. Souza for several and useful discussions.

REFERENCES

1. K. S. Miller and B. Ross, "An Introduction to the Fractional Calculus and Fractional Differential Equations," John Wiley & Sons, Inc., New York, 1993.

2. K. B. Oldham and J. Spanier, "The Fractional Calculus: Theory and Application of Differentiation and Integration to Arbitrary Order," Academic Press, New York, 1974.
3. R. L. Bagley and P. J. Torvik, "A Theoretical Basis for the Application of Fractional Calculus to Viscoelasticity," Journal of Rheology, Vol. 27, No. 3, 1983, pp. 201-210.doi:10.1122/1.549724
4. J. A. Tenreiro Machado, V. Kiryakova and F. Mainardi, "A Poster about the Recent History of Fractional Calculus," Fractional Calculus & Applied Analysis, Vol. 13, No. 3, 2010, pp. 329-334.
5. J. A. Tenreiro Machado, V. Kiryakova and F. Mainardi, "A Poster about the Old History of Fractional Calculus," Fractional Calculus & Applied Analysis, Vol. 13, No. 4, 2010, pp. 447-454.
6. J. A. Tenreiro Machado, V. Kiryakova and F. Mainardi, "Recent History of Fractional Calculus," Communications in Nonlinear Science and Numerical Simulation, Vol. 16, No. 3, 2011, pp. 1140-1153. doi:10.1016/j.cnsns.2010.05.027
7. K. Diethelm, "The Analysis of Fractional Differential Equations," Springer Verlag, Berlin, Heidelberg, 2010. doi:10.1007/978-3-642-14574-2
8. R. Hilfer, "Applications of Fractional Calculus in Physics," World Scientific, Singapore City, 2000.
9. A. A. Kilbas, H. M. Srivastava and J. J. Trujillo, "Theory and Applications of Fractional Differential Equations," Elsevier, Amsterdam, 2006.
10. I. Podlubny, "Fractional Differential Equations," Academic Press, San Diego, 1999.
11. S. G. Samko, A. A. Kilbas and O. I. Marichev, "Fractional ntegrals and Derivatives: Theory and Applications," Gordon and Breach Science Publishers, Amsterdam, 1993.
12. V. E. Tarasov, "Fractional Dynamics, Applications of Fractional Calculus to Dynamics of Particles, Fields and Media," Springer, Heidelberg, 2010.
13. V. E. Tarasov, "Fractional Vector Calculus and Fractional Maxwell's Equations," Annals of Physics, Vol. 323, No. 11, 2008, pp. 2756-2778. doi:10.1016/j.aop.2008.04.005
14. H. Vic Dannon, "The Fundamental Theorem of the Fractional Calculus and the Meaning of Fractional Derivatives," Gauge Institute Journal, Vol. 5, No. 1, 2009, pp. 1-26.
15. N. Heymans and I. Podlubny, "Physical Interpretation of Initial Conditions for Fractional Differential Equations with Riemann-Liouville Fractional Derivative," Rheologica Acta, Vol. 45, No. 5, 2006, pp. 765-772. doi:10.1007/s00397-005-0043-5
16. I. Podlubny, "Geometric and Physical Interpretation of Fractional Integral and Fractional Differentiation," Journal of Fractional Calculus & Applied Analysis, Vol. 5, No. 4, 2002, pp. 367-386.

17. A. Cabada and G. Wang, "Positive Solutions of Nonlinear Fractional Differential Equations with Integral Boundary Value Conditions," Journal of Mathematical Analysis and Applications, Vol. 389, No. 1, 2012, pp. 403-411. doi:10.1016/j.jmaa.2011.11.065

18. R. Figueiredo Camargo, "Fractional Calculus and Applications (in Portuguese)" Doctoral Thesis, UNICAMP, Campinas, 2009.

19. J. A. Tenreiro Machado, "Discrete-Time Fractional-Order Controllers," Journal of Fractional Calculus & Applied Analysis, Vol. 4, No. 1, 2001, pp. 47-66.

20. R. Figueiredo Camargo, E. Capelas de Oliveira and J. Vaz Jr., "On the Generalized Mittag-Leffler Function and Its Application in a Fractional Telegraph Equation," Mathematical Physsics, Analysis & Geometry, Vol. 15, No. 1, 2012, pp. 1-16. doi:10.1007/s11040-011-9100-8

21. F. Silva Costa and E. Capelas de Oliveira, "Fractional Wave-Diffusion Equation with Periodic Conditions," Journal of Mathematical Physics, Vol. 53, 2012, Article ID: 123520.doi:10.1063/1.4769270

22. H. Jafari and S. Momani, "Solving Fractional Diffusion and wave Equations by Modified Homotopy Perturbation Method," Physics Letters A, Vol. 370, No. 5-6, 2007, pp. 388-396.doi:10.1016/j.physleta.2007.05.118

23. E. Contharteze Grigoletto, "Fractional Differential Equations and the Mittag-Leffler Functions (in Portuguese)," Ph.D. Thesis, UNICAMP, Campinas, to Appear.

CITATION

1. E. Grigoletto and E. Oliveira, "Fractional Versions of the Fundamental Theorem of Calculus," Applied Mathematics, Vol. 4 No. 7A, 2013, pp. 23-33 doi: 10.4236/am.2013.47A006.

Differential Calculus for *P*-Norms of Complex-Valued Vector Functions with Applications

L. Kohaupt
Prager Strasse 9, D-10779 Berlin, Germany

ABSTRACT

For complex-valued n-dimensional vector functions $t \mapsto s(t)$, supposed to be sufficiently smooth, the differentiability properties of the mapping $t \mapsto \|s(t)\|_p$ at every point $t = t_0 \in \mathbb{R}_0^+ := \{t \in \mathbb{R} \mid t \geq 0\}$ are investigated, where $\|\cdot\|_p$ is the usual vector norm in $\mathbb{C}^n$ resp. $\mathbb{R}^n$, for $p \in [1, \infty]$. Moreover, formulae for the first three right derivatives $D_+^k \|s(t)\|_p, k = 1,2,3$ are determined. These formulae are applied to vibration problems by computing the best upper bounds on $\|s(t)\|_p$ in certain classes of bounds. These results cannot be obtained by the methods used so far. The systematic use of the differential calculus for vector norms, as done here for the first time, could lead to major advances also in other branches of mathematics and other sciences.

INTRODUCTION

First, this paper studies systematically the differentiability properties for p-norms of complex-valued vector functions, that is, it develops a pertinent differential calculus. Second, the results are applied to compute upper bounds on the norm of the solution of vibration problems (free and force-excited vibrations); the obtained upper bounds are the best possible ones in the considered classes of bounds, a result which in general cannot be achieved by the methods used so far.

In [11], the author has generalized the concept of logarithmic derivative by proving that for the fundamental matrix $\Phi(t)=e^{At}$ (with a complex $n \times n$ matrix A) the function $t \mapsto \|\Phi(t)\|_p$ is real analytic on some neighborhood $[t_0, t_0+\Delta t_0]$ (with $t_0 \in \mathbb{R}_0^+, \Delta t_0 > 0$ sufficiently small, and $p \in \{1,2,\infty\}$). Also, formulae for the right derivatives $D_+^k \|\Phi(t)\|_p, k = 1,2$ are determined in [11]. With these results, the best upper bounds such that $\|\Phi(t)\|_p \leq M_{\varepsilon,p}e^{(\alpha+\varepsilon)t}$ have been computed there, where ε was set equal to eps=2^{-52} (the machine precision in MATLAB) and where α is the spectral abscissa of the matrix A.

Since the solution of the initial-value problem $\dot{x}(t) = Ax(t), x(0) = x_0$ is given by $x(t)=\Phi(t)x_0$, an upper bound on $x(t)$ can be obtained by the estimate $\|x(t)\|_p \leq \|\Phi(t)\|_p \|x_0\|_p \leq X_{\varepsilon,p}e^{(\alpha+\varepsilon)t}$ with $X_{\varepsilon,p} = M_{\varepsilon,p} \|x_0\|_p$. This constant $X_{\varepsilon,p}$ is, however, not optimal. To obtain the minimal value of $X_{\varepsilon,p}$, one has to develop a differential calculus for p-norms of complex-valued vector functions just as for the p-norms of the fundamental matrix. This is the main subject of the present paper. Apart from investigating in the applications a problem of free vibration described by $\dot{x}(t) = Ax(t), x(0) = x_0$, we also study problems for sinoidal force-excited vibrations with constant and linearly increasing amplitude described by an equation of the form $\dot{x}(t) = Ax(t) + g(t), x(0) = x_0$ with $g(t) = g_0 \sin \omega t$ resp. $g(t) = g_0 t \sin \omega t$. Also, for these problems the best upper bounds in the considered classes of bounds are computed. Further, we apply the second and third right derivative of a function $\|x(\cdot)\|_p$ to determine its inflexion points in a certain region.

The paper is structured as follows. In Section 2, regularity properties of the mapping $t \mapsto \| s(t) \| p, \in [1,\infty]$, at every point $t = t_0 \in \mathbb{R}_0^+$ are stated, where $s(\cdot)$ is a complex-valued n-dimensional vector function, which is supposed to be m times continuously differentiable for $t \geq 0$. In Section 3, formulae for the right derivatives can be found. Section 4 is the application part. Here, the results from the previous sections are applied to vibration problems with and without force excitation to obtain the best upper bounds in certain classes of bounds. Further, it is pointed out that the developed differential calculus can be used in the discussion of the functions $t \mapsto \|x(t)\|_p$, e.g. for the determination of inflexion points. In Section 5, some concluding remarks are made. Finally, Section 6 is an appendix compiling some useful auxiliary means needed in Section 4; the applied method of 6.1 seems to be new (cf. [13, pp. 197–198]).

LOCAL REGULARITY

First, as a preparation we collect some properties of complex vector functions (Lemma 1 and Lemma 2). Then, the main lemma on the local regularity (Lemma 3) follows, which plays a fundamental theoretical role in this paper.

Let $a, b \in \mathbb{R}, a \neq b$, and define $\langle a, b \rangle := [\min(a,b), \max(a,b)]$.

Lemma 1

Let $s : \langle t_0, t \rangle \to \mathbb{C}^n$ be an n-dimensional vector function that is m times continuously differentiable, and denote by $D^k s(t_0)$ the kth derivative of s at $t = t_0$. Then,

$$s(t) = \sum_{k=0}^{m} D^k s(t_0) \frac{(t - t_0)^k}{k!} + R(t),$$

(1)

where $R : \langle t_0, t \rangle \to \mathbb{C}^n$ is m times continuously differentiable with the properties $D^k R(t_0) = 0, k = 0, 1, \ldots, m$ and $R(t) = o((t-t_0)^m)$ as $t \to t_0$.

Proof

Case 1: $s : \langle t_0, t \rangle \to \mathbb{R}$. That $R : \langle t_0, t \rangle \to \mathbb{R}$ is m times continuously differentiable and $D^k R(t_0) = 0, k = 0, 1, \ldots, m$ follows directly from the representation

$$R(t) = s(t) - \sum_{k=0}^{m} D^k s(t_0) \frac{(t - t_0)^k}{k!}.$$

(2)

Further, according to [6, Theorem 61.1 (Taylor's Theorem), p. 355], it follows that there exists a number $\vartheta \in (0,1)$ such that

$$s(t) = \sum_{k=0}^{m-1} D^k s(t_0) \frac{(t - t_0)^k}{k!} + D^m s(t_0 + \theta(t - t_0)) \frac{(t - t_0)^m}{m!}.$$

(3)

(For this, we need only the following hypotheses: s is $(m-1)$ times continuously differentiable, and $D^m s$ exists in int $\langle t_0, t \rangle$)

Since $s \in C^m \langle t_0, t \rangle$, the last relation can be cast into the form

$$s(t) = \sum_{k=0}^{m} D^k s(t_0) \frac{(t - t_0)^k}{k!} + R(t),$$

(4)

where

$$R(t) = [D^m s(t_0 + \theta(t - t_0)) - D^m s(t_0)] \frac{(t - t_0)^m}{m!}.$$

(5)

Thus,

$$R(t) = o((t - t_0)^m)$$

(6)

since

$$D^m s(t_0 + \theta(t - t_0)) - D^m s(t_0) = o(1)$$

(7)

as $t \to t_0$.

Case 2: $s : \langle t_0, t \rangle \to \mathbb{C}$. From Case 1, the proof follows for the real and imaginary parts of s, where instead of a value $\vartheta \in (0,1)$ two (in general different) values $\vartheta_{\text{real}}, \vartheta_{\text{imag}} \in (0,1)$ for the real and imaginary parts enter, respectively.

Case 3: $s : \langle t_0, t \rangle \to \mathbb{C}^n, n > 1$. Here, Case 2 is applied to each component s_i of $s = [s_1, \ldots, s_n]^T$, where instead of a single pair $\vartheta_{real}, \vartheta_{imag} \in (0,1)$ pairs $\theta_{real}^{(i)}, \theta_{imag}^{(i)} \in (0,1)$ enter, respectively.

Lemma 2

Let the n-dimensional vector function $s : \langle t_0, t \rangle \to \mathbb{C}^n$ be of the form

$$s(t) = \sum_{k=0}^{m} \lambda^{(k)} \frac{(t - t_0)^k}{k!} + R(t),$$

$$(8)$$

where $R : \langle t_0, t \rangle \to \mathbb{C}^n$ is supposed to be m times continuously differentiable with

$$D^k R(t_0) = 0, \quad k = 0, 1, \ldots, m.$$

$$(9)$$

Then, $s : \langle t_0, t \rangle \to \mathbb{C}^n$ is m times continuously differentiable with the property

$$\lambda^{(k)} = D^k s(t_0), \quad k = 0, 1, \ldots, m$$

$$(10)$$

and

$$R(t) = o((t - t_0)^m)$$

$$(11)$$

as $t \to t_0$.

Proof

It is clear that $s : \langle t_0, t \rangle \to \mathbb{C}^n$ is m times continuously differentiable and that $\lambda^{(k)} = D^k s(t_0), k = 0, 1, \ldots, m$. The rest then follows from Lemma 1.

Lemma 3

Let $s : \langle t_0, t \rangle \to \mathbb{C}^n$ be an n-dimensional complex-valued vector function that is m times continuously differentiable, and let $t_0 \in \mathbb{R}_0^+$. Further, let $p \in [1,\infty]$, and in the case $p=\infty$ suppose additionally that each two components of s are either identical or intersect each other at most finitely often near t_0.

Then, there exists a number $\Delta t_0 > 0$ and a function $t \mapsto \hat{s}(t)$, which is real and m times continuously differentiable on $[t_0, t_0 + \Delta t_0]$ such that $t \mapsto \hat{s}(t) = \|s(t)\|_\mu$ for all $t \in [t_0, t_0 + \Delta t_0]$.

Proof

Let $s = [s_1, \ldots, s_n]^T$. Then,

$$|s_i(t)| = \sqrt{[\operatorname{Re} s_i(t)]^2 + [\operatorname{Im} s_i(t)]^2}, \quad i = 1, \ldots, n \tag{12}$$

for $t \geq 0$. By Lemma 1, because of $\operatorname{Re} D^k s_i(t_0) = D^k \operatorname{Re} s_i(t_0)$ and $\operatorname{Im} D^k s_i(t_0) = D^k \operatorname{Im} s_i(t_0)$ we have

$$\operatorname{Re} s_i(t) = \sum_{k=0}^{m} D^k \operatorname{Re} s_i(t_0) \frac{(t - t_0)^k}{k!} + \operatorname{Re} R_i(t) \tag{13}$$

$$\operatorname{Im} s_i(t) = \sum_{k=0}^{m} D^k \operatorname{Im} s_i(t_0) \frac{(t - t_0)^k}{k!} + \operatorname{Im} R_i(t) \tag{14}$$

$t \geq t_0$, where $\operatorname{Re} R_i, \operatorname{Im} R_i : \langle t_0, t \rangle \to \mathbb{R}$ are m times continuously differentiable with the properties $D^k \operatorname{Re} R_i(t_0) = 0, D^k \operatorname{Im} R_i(t_0) = 0, k = 0, 1, \ldots m$. Therefore, if we choose $\Delta t_1 > 0$ sufficiently small, also the function $t \mapsto |s_i(t)|$ is m times continuously differentiable and can be written in the form

$$|s_i(t)| = \sum_{k=0}^{m} \lambda_i^{(k)} \frac{(t - t_0)^k}{k!} + r_i(t), \quad t_0 \leq t \leq t_0 + \Delta t_1, \tag{15}$$

where $\lambda_i^{(k)} = D^k |s_i(t_0)| := D_+^k |s_i(t_0)|, k = 0, 1, \ldots, m$, where $r_i : [t_0, t_0 + \Delta t_1] \to \mathbb{R}$ is m times continuously differentiable with $D^k r_i(t_0) := D_+^k r_i(t_0) = 0, 1, \ldots, m$ for $i = 1, \ldots, n$, and where $r_i(t) = ((t - t_0)^m), i = 1, \ldots, n$.

p=∞: we have

$$\|s(t)\|_\infty = \max_{i=1,\dots,n} |s_i(t)|, \quad t \geq 0.$$

(16)

Let $I_{-1} := \{1,\dots,n\}$ and I_k be the set of all indices $i_k \in I_{k-1}$ where $\lambda_i^{(k)}$ attains its maximum, i.e.

$$I_k := \left\{ i_k \in I_{k-1} \, \middle| \, \lambda_{i_k}^{(k)} = \max_{i \in I_{k-1}} \lambda_i^{(k)} \right\}, \quad k = 0, 1, \dots, m.$$

(17)

Then, by a repeated application of [10, Lemma 2.1] (with t replaced by t−t_0), we obtain a number Δt_2 with $0 < \Delta t_2 \leq \Delta t_1$ such that

$$\|s(t)\|_\infty = \max_{i=1,\dots,n} |s_i(t)| = \max_{i \in I_{-1}} \left\{ \sum_{k=0}^{m} \lambda_i^{(k)} \frac{(t - t_0)^k}{k!} + r_i(t) \right\}$$

$$= \sum_{k=0}^{m-1} \lambda_{i_k}^{(k)} \frac{(t - t_0)^k}{k!} + (t - t_0)^m \max_{i \in I_{m-1}} \left\{ \frac{\lambda_i^{(m)}}{m!} + r_i^{(m)}(t) \right\},$$

(18)

where

$$r_i^{(m)}(t) = \begin{cases} \dfrac{r_i(t)}{(t - t_0)^m}, & t_0 < t \leq t_0 + \Delta t_2 \\[2mm] \lim_{t \to t_0} \dfrac{r_i(t)}{(t - t_0)^m} = 0, & t = t_0. \end{cases}$$

Thus,

$$\|s(t)\|_\infty = \sum_{k=0}^{m} \lambda_{i_k}^{(k)} \frac{(t - t_0)^k}{k!} + \max_{i \in I_m} r_i(t), \quad t_0 \leq t \leq t_0 + \Delta t_2.$$

(19)

Now, because of the additional hypothesis for p=∞, there exists an index set $I_{m+1}^{(r)} \subset I_m$ and a number Δt_3 with $0 < \Delta t_3 \leq \Delta t_2$ such that

$$r^{(r)}_{i_{m+1}}(t) = \max_{i \in I_m} r_i(t), \quad i^{(r)}_{m+1} \in I^{(r)}_{m+1}, \quad t_0 \leqslant t \leqslant t_0 + \Delta t_3.$$

Choose any index $i^{(r)}_{m+1} \in I^{(r)}_{m+1}$. Then,

$$\|s(t)\|_\infty = \sum_{k=0}^m \lambda^{(k)}_{i_k} \frac{(t-t_0)^k}{k!} + r^{(r)}_{i_{m+1}}(t), \quad t_0 \leqslant t \leqslant t_0 + \Delta t_3.$$

So, with $\Delta t_0 := \Delta t_3$ and $\hat{s}(t) := \sum_{k=0}^m \lambda^{(k)}_{ik} \frac{(t-t_0)^k}{k!} + r^{(r)}_{i_{m+1}}(t)$ the assertion follows.

$p \in [1, \infty)$: Due to (15), there exists a number $\Delta t_2 > 0$ with $\Delta t_2 \leq \Delta t_1$ such that

$$\|s(t)\|_p = \left(\sum_{i=1}^n |s_i(t)|^p \right)^{1/p} = \sum_{k=0}^m \lambda_k \frac{(t-t_0)^k}{k!} + r(t), \quad t_0 \leqslant t \leqslant t_0 + \Delta t_1,$$

$$(20)$$

where λ_k and $r(t)$ depend on p. So, the assertion follows with $\Delta t_0 := \Delta t_2$ and $\hat{s}(t) := \sum_{k=0}^m \lambda_k (t-t_0)^k / k! + r(t)$, where $r : [t_0, t_0 + \Delta t_2] \to \mathbb{R}$ is m times continuously differentiable and $D^k r(t_0) = 0, k = 0, 1, ..., m$ as well as $r(t) = o((t-t_0)m)$. The details are left to the reader.

Supplement 4

If $s(\cdot)$ in Lemma 3 is analytic for $t \geq 0$ (or in a neighborhood of the considered point $t_0 \in \mathbb{R}^+_0$), then the additional hypothesis for $p = \infty$ can be dropped.

Proof

If $s(\cdot)$ is analytic for $t \geq 0$, it is analytic in a neighborhood of $t_0 \in \mathbb{R}^+_0$. Let $\Delta t_1 > 0$ be a sufficiently small positive number such that $|s_i(t)|$ are analytic in $[t_0, t_0 + \Delta t_1]$. Suppose that there exist two indices $j, k \in \{1, ..., n\}$ such that $|s_j(t)|$ and $|s_k(t)|$ intersect infinitely often in the interval $[t_0, t_0 + \Delta t_1]$. The intersection points have a limit point in $[t_0, t_0 + \Delta t_1]$ since this interval is compact in $\mathbb{R}$. Then, according to the identity theorem for analytic functions (cf. [9, Part I, 21, p. 87]), we have $|s_j(t)| = |s_k(t)|, t \in [t_0, t_0 + \Delta t_2]$. Now consider the case when $|s_j(t)|$ and $|s_k(t)|$ intersect at most finitely often. Then, there exists a number Δt_2 with $0 < \Delta t_2 \leq \Delta t_1$ such that $|s_j(t)| \leq |s_k(t)|, t \in [t_0, t_0 + \Delta t_2]$ or $|s_j(t)| \geq |s_k(t)|, t \in [t_0, t_0 + \Delta t_2]$. Considering in this way all pairs $j, k \in \{1, \ldots, n\}$, we conclude that there exists a number Δt_3 with $0 < \Delta t_3 \leq \Delta t_2$ and an index set $I_N \subset I_{-1} = \{1, \ldots, n\}$ such that

$$\max_{i \in \{1, \ldots, n\}} |s_i(t)| = |s_j(t)|, \quad j \in I_N \ t \in [t_0, t_0 + \Delta t_3],$$

where $|s_j(t)| = |s_k(t)|, j, k \in I_N, t \in [t_0, t_0 + \Delta t_3]$. (In practise, however, one does not know the index set IN in advance) For the series expansion of $\|s(t)\|_\infty$, one can use the finite sequence

$$I_{-1} \supset I_0 \supset I_1 \supset \cdots \supset I_N = I_i, \quad i \geqslant N$$

$$(21)$$

in the proof of Lemma 3, for $m = \infty$. So, one obtains the representation

$$\|s(t)\|_\infty = \sum_{k=0}^{N-1} \lambda_{i_k}^{(k)} \frac{(t - t_0)^k}{k!} + \sum_{k=N}^{\infty} \lambda_{i_N}^{(k)} \frac{(t - t_0)^k}{k!},$$

$$(22)$$

for any indices $i_k \in I_k, k = 0, 1, \ldots, N$ and $t \in [t_0, t_0 + \Delta t_0]$ for a sufficiently small number $\Delta t_0 > 0$. Of course, one could also write

$$\|s(t)\|_\infty = \sum_{k=0}^{\infty} \lambda_{i_N}^{(k)} \frac{(t - t_0)^k}{k!}$$

$$(23)$$

for $t \in [t_0, t_0 + \Delta t_0]$ since $\lambda_{i_k}^{(k)} = \lambda_{i_N}^{(k)}, k = 0, 1, \ldots, N$.

Remark
In the application part, the vector functions $x(\cdot)$ are analytic so that Supplement 4 applies.

FORMULAE FOR THE RIGHT DERIVATIVES

According to Lemma 3, all right derivatives $D_+^k \, || \, s(t) \, ||_p, t \in \mathbb{R}_0^+, p \in [1,\infty]$, exist for sufficiently smooth vector functions. In the case of complex functions, we restrict ourselves to $k \in \{1, 2, 3\}$.

We begin with the special cases $p \in \{1, 2, \infty\}$. Even though the cases p $\in \{1, 2\}$ can be deduced from the general case $1 \leq p < \infty$, they are stated here separately because they take special, much simpler forms. The special cases are followed by the general case $1 \leq p < \infty$. We leave it to the reader to derive the cases $p \in \{1, 2\}$ from the general case.

$p = \infty$: Complex vector function $s \in C^m(\mathbb{R}_0^+, \mathbb{C}^n)$.

Let $t_0 \in \mathbb{R}_0^+$ and $C^m(\mathbb{R}_0^+, \mathbb{C}^n)$ be the space of functions $s(\cdot)$ defined on $\mathbb{R}_0^+$ with values in $\mathbb{C}^n$ such that $s(\cdot)$ is m times continuously differentiable. If $m = 3$, then

$$s(t) = x + (t - t_0)y + \frac{(t - t_0)^2}{2!}z + \frac{(t - t_0)^3}{3!}u + r(t), \quad t \geq t_0,$$

$$(24)$$

where

$$x = s(t_0)$$
$$y = Ds(t_0)$$
$$z = D^2 s(t_0)$$
$$u = D^3 s(t_0)$$

and

$$r(t) = o((t - t_0)^m).$$

With these vectors, define the following functionals for $I \in \{1,\dots,n\}$:

$$S_i^{(0)} := |x_i|,$$

$$(25)$$

$$S_i^{(1)} := \begin{cases} \dfrac{\operatorname{Re} x_i \operatorname{Re} y_i + \operatorname{Im} x_i \operatorname{Im} y_i}{|x_i|}, & x_i \neq 0, \\[2mm] |y_i|, & x_i = 0, \end{cases}$$

(26)

$$S_i^{(2)} := \begin{cases} \dfrac{|y_i|^2 + \operatorname{Re} x_i \operatorname{Re} z_i + \operatorname{Im} x_i \operatorname{Im} z_i}{|x_i|} \\[3mm] \quad - \dfrac{[\operatorname{Re} x_i \operatorname{Re} y_i + \operatorname{Im} x_i \operatorname{Im} y_i]^2}{|x_i|^3}, & x_i \neq 0 \\[4mm] \dfrac{\operatorname{Re} y_i \operatorname{Re} z_i + \operatorname{Im} y_i \operatorname{Im} z_i}{|y_i|}, & x_i = 0, \ y_i \neq 0 \\[3mm] |z_i|, & x_i = 0, \ y_i = 0 \end{cases}$$

(27)

$$S_i^{(3)} := \begin{cases} \dfrac{[\operatorname{Re} x_i \operatorname{Re} u_i + \operatorname{Im} x_i \operatorname{Im} u_i] + 3[\operatorname{Re} y_i \operatorname{Re} z_i + \operatorname{Im} y_i \operatorname{Im} z_i]}{|x_i|} \\[3mm] \quad - \dfrac{3[\operatorname{Re} x_i \operatorname{Re} y_i + \operatorname{Im} x_i \operatorname{Im} y_i][|y_i|^2 + \operatorname{Re} x_i \operatorname{Re} z_i + \operatorname{Im} x_i \operatorname{Im} z_i]}{|x_i|^3} \\[3mm] \quad + \dfrac{3[\operatorname{Re} x_i \operatorname{Re} y_i + \operatorname{Im} x_i \operatorname{Im} y_i]^3}{|x_i|^5}, & x_i \neq 0 \\[4mm] \dfrac{3|z_i|^2 + 4[\operatorname{Re} y_i \operatorname{Re} u_i + \operatorname{Im} y_i \operatorname{Im} u_i]}{4|y_i|} \\[3mm] \quad - \dfrac{3[\operatorname{Re} y_i \operatorname{Re} z_i + \operatorname{Im} y_i \operatorname{Im} z_i]^2}{4|y_i|^3}, & x_i = 0, \ y_i \neq 0 \\[4mm] \dfrac{\operatorname{Re} z_i \operatorname{Re} u_i + \operatorname{Im} z_i \operatorname{Im} u_i}{|z_i|}, & x_i = 0, \ y_i = 0, \ z_i \neq 0 \\[3mm] |u_i|, & x_i = 0, \ y_i = 0, \ z_i = 0 \end{cases}$$

(28)

Hereby, the next theorem can be proved.

Theorem 5

($p = \infty$) Let $s : \mathbb{R}_0^+ \to \mathbb{C}^n$ be an n-dimensional complex-valued vector function that is m=3 times continuously differentiable, and let $t_0 \in \mathbb{R}_0^+$. Suppose additionally that each two components of s are either identical or intersect each other at most finitely often near t_0. Further, let $I_{-1} = \{1,\ldots,n\}$ and I_0 be the index set where S_i^0 attains its maximum,

$$I_0 := \left\{ i_0 \in I_{-1} \mid S_{i_0}^{(0)} = \max_{i \in I_{-1}} S_i^{(0)} \right\}.$$

(29)

Similarly, let

$$I_k := \left\{ i_k \in I_{k-1} \mid S_{i_k}^{(k)} = \max_{i \in I_{k-1}} S_i^{(k)} \right\},$$

(30)

k=1, 2, 3. Then,

$$\|s(t_0)\|_\infty = \max_{i \in I_{-1}} S_i^{(0)}$$

(31)

$$D_+^1 \|s(t_0)\|_\infty = \max_{i \in I_0} S_i^{(1)}$$

(32)

$$D_+^2 \|s(t_0)\|_\infty = \max_{i \in I_1} S_i^{(2)}$$

(33)

$$D_+^3 \|s(t_0)\|_\infty = \max_{i \in I_2} S_i^{(3)}$$

(34)

Proof
We have

$$\|s(t)\|_\infty = \left\| x + (t - t_0)\,y + \frac{(t - t_0)^2}{2!}\,z + \frac{(t - t_0)^3}{3!}\,u + r(t) \right\|_\infty$$

$$= \max_{i=1,\ldots,n} \left| x_i + (t - t_0)\,y_i + \frac{(t - t_0)^2}{2!}\,z_i + \frac{(t - t_0)^3}{3!}\,u_i + r_i(t) \right|$$

$$= \max_{i=1,\ldots,n} |S_i|,$$

(35)

where

$$S_i = x_i + (t - t_0)\,y_i + \frac{(t - t_0)^2}{2!}\,z_i + \frac{(t - t_0)^3}{3!}\,u_i + r_i(t),$$

i=1,...,n. For sufficiently small t≥t$_0$,

$$|S_i| = S_i^{(0)} + S_i^{(1)}(t - t_0) + S_i^{(2)}\frac{(t - t_0)^2}{2!} + S_i^{(3)}\frac{(t - t_0)^3}{3!} + p_i(t),$$
(36)

where $p_i(t) = o((t - t_0)^3), i = 1,...,n$. Formulae, and (31)-(34) follow in a similar way as in the proof of Lemma 3 for p=∞, which in turn relies on [10, Lemma 2.1].

p=∞: Real vector function $s \in C^m(\mathbb{R}_0^+, \mathbb{R}^n)$.

In this case, a unified formula for all right derivatives of $\|s(\cdot)\|_\infty$ exists. To show this, define the following sign functionals for $I \in \{1,...,n\}$:

$$s_i^{(0)} := \mathrm{sgn}[s_i(t_0)],$$
(37)

$$s_i^{(1)} := \begin{cases} \mathrm{sgn}[s_i(t_0)], & s_i(t_0) \neq 0, \\ \mathrm{sgn}[Ds_i(t_0)], & s_i(t_0) = 0, \end{cases}$$
(38)

$$s_i^{(m)} := \begin{cases} \mathrm{sgn}[s_i(t_0)], & s_i(t_0) \neq 0, \\ \mathrm{sgn}[Ds_i(t_0)], & s_i(t_0) = 0, \ Ds_i(t_0) \neq 0, \\ \vdots \\ \mathrm{sgn}[D^m s_i(t_0)], & D^k s_i(t_0) = 0, \ k = 0, 1, \ldots, m - 1 \end{cases}$$
(39)

or briefly,

$$s_i^{(k)} = \begin{cases} s_i^{(k-1)}, & s_i^{(k-1)} \neq 0, \\ \mathrm{sgn}[D^k s_i(t_0)], & s_i^{(k-1)} = 0, \end{cases}$$
(40)

i=1,...,n; k=1,...,m. With these sign functionals, define the further functionals

$$S_i^{(k)} := s_i^{(k)} \cdot D^k s_i(t_0), \quad i = 1, \ldots, n; \ k = 0, 1, \ldots, m.$$
(41)

Then, the right derivatives for real vector functions read as follows.

Theorem 6 (p=∞, real vector function)

Let $s: \mathbb{R}_0^+ \to \mathbb{R}^n$ be an n-dimensional real-valued vector function that is m times continuously differentiable, and let $t_0 \in \mathbb{R}_0^+$. Suppose additionally that each two components of s are either identical or intersect each other at most finitely often near t_0. Further, let $I_{-1}=\{1,\dots,n\}$ and I_k be the set of all indices $i_k \in I_{k-1}$ where $S_i^{(k)}$ from (41) attains its maximum, i.e.

$$I_k := \left\{ i_k \in I_{k-1} \mid S_{i_k}^{(k)} = \max_{i \in I_{k-1}} S_i^{(k)} \right\},$$

$$\tag{42}$$

k=1,...,m. Then, the right derivatives of $t \mapsto \|s(t)\|_\infty$ at $t=t_0 \geq 0$ are given by

$$D_+^k \|s(t_0)\|_\infty = \max_{i \in I_{k-1}} S_i^{(k)}, \quad k = 1, \dots, m.$$

$$\tag{43}$$

The proof is left to the reader.

p=1: Complex vector function $s \in C^m(\mathbb{R}_0^+, \mathbb{C}^n)$.

Theorem 7 p=1

Let $s: \mathbb{R}_0^+ \to \mathbb{C}^n$ be an n-dimensional complex-valued vector function that is m = 3 times continuously differentiable, and let $t_0 \in \mathbb{R}_0^+$.

Then,

$$\|s(t_0)\|_1 = \sum_{i=1}^{n} S_i^{(0)},$$

$$\tag{44}$$

$$D_+^1 \|s(t_0)\|_1 = \sum_{i=1}^{n} S_i^{(1)},$$

$$\tag{45}$$

$$D_+^2 \|s(t_0)\|_1 = \sum_{i=1}^{n} S_i^{(2)},$$

$$\tag{46}$$

$$D_+^3 \|s(t_0)\|_1 = \sum_{i=1}^{n} S_i^{(3)},$$

$$(47)$$

where $S_i^{(0)} - S_i^{(3)}$ are defined by (25)–(28).

Proof

The proof follows from

$$\|s(t)\|_1 = \left\| x + (t - t_0)\,y + \frac{(t - t_0)^2}{2!}\,z + \frac{(t - t_0)^3}{3!}\,u + r(t) \right\|_1$$

$$= \sum_{i=1}^{n} \left| x_i + (t - t_0)\,y_i + \frac{(t - t_0)^2}{2!}\,z_i + \frac{(t - t_0)^3}{3!}\,u_i + r_i(t) \right|$$

$$= \sum_{i=1}^{n} |S_i|,$$

$$(48)$$

where $|S_i|$ is given in (36).

$p=2$: Vector function $s \in C^m(\mathbb{R}_0^+, \mathbb{C}^n)$ resp. $s \in C^m(\mathbb{R}_0^+, \mathbb{R}^n)$.

We treat only the complex case since the real case is a special case of the complex one, and we start with the expansion (24). Define the following functionals:

$$S^{(0)} := \|x\|_2,$$

$$(49)$$

$$S^{(1)} := \begin{cases} \dfrac{\text{Re}(x, y)}{\|x\|_2}, & x \neq 0 \\[4mm] \|y\|_2, & x = 0 \end{cases}$$

$$(50)$$

$$S^{(2)} := \begin{cases} \dfrac{\|y\|_2^2 + \mathrm{Re}(x,z)}{\|x\|_2} - \dfrac{[\mathrm{Re}(x,y)]^2}{\|x\|_2^3}, & x \neq 0 \\[3mm] \dfrac{\mathrm{Re}(y,z)}{\|y\|_2}, & x = 0, \ y \neq 0 \\[3mm] \|z\|_2, & x = 0, \ y = 0 \end{cases}$$

(51)

$$S^{(3)} := \begin{cases} \dfrac{\mathrm{Re}(x,u) + 3\,\mathrm{Re}(y,z)}{\|x\|_2} - \dfrac{3\,\mathrm{Re}(x,y)[\|y\|_2^2 + \mathrm{Re}(x,z)]}{\|x\|_2^3} \\[3mm] \quad + \dfrac{3[\mathrm{Re}(x,y)]^3}{\|x\|_2^5}, & x \neq 0 \\[3mm] \dfrac{3\|z\|_2^2 + 4\,\mathrm{Re}(y,u)}{4\|y\|_2} - \dfrac{3[\mathrm{Re}(y,z)]^2}{4\|y\|_2^3}, & x = 0, \ y \neq 0 \\[3mm] \dfrac{\mathrm{Re}(z,u)}{\|z\|_2}, & x = 0, \ y = 0, \ z \neq 0 \\[3mm] \|u\|_2, & x = 0, \ y = 0, \ z = 0 \end{cases}$$

(52)

Then, we obtain

Theorem 8 p=2

Let $\mathbb{R}_0^+ \to \mathbb{C}^n$ resp. $s : \mathbb{R}_0^+ \to \mathbb{R}^n$ be an n-dimensional vector function that ism=3 times continuously differentiable, and let $t_0 \in \mathbb{R}_0^+$.

Then,

$$\|s(t_0)\|_2 = S^{(0)},$$

(53)

$$D_+^1 \|s(t_0)\|_2 = S^{(1)},$$

(54)

$$D_+^2 \|s(t_0)\|_2 = S^{(2)},$$

(55)

$$D_+^3 \|s(t_0)\|_2 = S^{(3)},$$

(56)

where $S^{(0)}$–$S^{(3)}$ are defined by (49)–(52).

Proof

For sufficiently small $\Delta t_0 > 0$, we obtain the expansion

$$\|s(t)\|_2 = S^{(0)} + S^{(1)}(t - t_0) + S^{(2)}\frac{(t - t_0)^2}{2!} + S^{(3)}\frac{(t - t_0)^3}{3!} + r(t), \tag{57}$$

where $r(t) = o((t - t_0)^m)$ for $t \in [t_0, t_0 + \Delta t_0]$.

Remark

The expressions of $D^k + \|s(t_0)\|_\infty$ and $D^k + \|s(t_0)\|_1$ on the one hand and $D^k + \|s(t_0)\|_2$ on the other hand look rather different. This can be changed by introducing a notation involving the components of the vectors resp. the scalar product of numbers. We demonstrate this for $D^k + \|s(t_0)\|_\infty$ and $D^k + \|s(t_0)\|_2$ in Appendix A, Section 4.2 resp. 4.3.

General case $1 \leq p < \infty$: Complex vector function $s \in C^m(\mathbb{R}_0^+, \mathbb{C}^n)$.

Let $S_i^{(0)}, S_i^{(1)}, S_i^{(2)}, S_i^{(3)}, i = 1, \ldots, n$ be given by (25)-(28) Hereby define the following functionals:

$$S^{(0,p)} := \left(\sum_{i=1}^n (S_i^{(0)})^p\right)^{1/p}, \tag{58}$$

$$S^{(1,p)} := \begin{cases} \dfrac{\sum_{i=1}^n (S_i^{(0)})^{p-1} S_i^{(1)}}{(\sum_{i=1}^n (S_i^{(0)})^p)^{1-1/p}}, & \left(\sum_{i=1}^n (S_i^{(0)})^p\right)^{1/p} \neq 0 \\[2ex] \left(\sum_{i=1}^n (S_i^{(1)})^p\right)^{1/p}, & \left(\sum_{i=1}^n (S_i^{(0)})^p\right)^{1/p} = 0 \end{cases} \tag{59}$$

$$S^{(2,p)} := \begin{cases} \dfrac{\sum_{i=1}^{n}(S_i^{(0)})^{p-1}S_i^{(2)} + (p-1)\sum_{i=1}^{n}(S_i^{(0)})^{p-2}(S_i^{(1)})^2}{(\sum_{i=1}^{n}(S_i^{(0)})^p)^{1-1/p}} \\[2mm] \quad + \dfrac{(1-p)[\sum_{i=1}^{n}(S_i^{(0)})^{p-1}S_i^{(1)}]^2}{(\sum_{i=1}^{n}(S_i^{(0)})^p)^{2-1/p}}, \quad \left(\sum_{i=1}^{n}(S_i^{(0)})^p\right)^{1/p} \neq 0 \\[4mm] \dfrac{\sum_{i=1}^{n}(S_i^{(1)})^{p-1}S_i^{(2)}}{(\sum_{i=1}^{n}(S_i^{(1)})^p)^{1-1/p}}, \quad \left(\sum_{i=1}^{n}(S_i^{(0)})^p\right)^{1/p} = 0, \quad \left(\sum_{i=1}^{n}(S_i^{(1)})^p\right)^{1/p} \neq 0 \\[4mm] \left(\sum_{i=1}^{n}(S_i^{(2)})^p\right)^{1/p}, \quad \left(\sum_{i=1}^{n}(S_i^{(0)})^p\right)^{1/p} = 0, \quad \left(\sum_{i=1}^{n}(S_i^{(1)})^p\right)^{1/p} = 0 \end{cases} \tag{60}$$

$$S^{(3,p)} := \begin{cases} \dfrac{\sum_{i=1}^{n}(S_i^{(0)})^{p-1}S_i^{(3)}+3(p-1)\sum_{i=1}^{n}(S_i^{(0)})^{p-2}S_i^{(1)}S_i^{(2)}+(p-1)(p-2)\sum_{i=1}^{n}(S_i^{(0)})^{p-3}(S_i^{(1)})^3}{(\sum_{i=1}^{n}(S_i^{(0)})^p)^{1-1/p}} \\[2mm] \quad + \dfrac{3(1-p)[\sum_{i=1}^{n}(S_i^{(0)})^{p-1}S_i^{(1)}][\sum_{i=1}^{n}(S_i^{(0)})^{p-1}S_i^{(2)} + (p-1)\sum_{i=1}^{n}(S_i^{(0)})^{p-2}(S_i^{(1)})^2]}{(\sum_{i=1}^{n}(S_i^{(0)})^p)^{2-1/p}} \\[2mm] \quad + \dfrac{(1-p)(1-2p)[\sum_{i=1}^{n}(S_i^{(0)})^{p-1}S_i^{(1)}]^3}{(\sum_{i=1}^{n}(S_i^{(0)})^p)^{3-1/p}}, \quad \left(\sum_{i=1}^{n}(S_i^{(0)})^p\right)^{1/p} \neq 0 \\[4mm] \dfrac{4\sum_{i=1}^{n}(S_i^{(1)})^{p-1}S_i^{(3)} + 3(p-1)\sum_{i=1}^{n}(S_i^{(1)})^{p-2}(S_i^{(2)})^2}{4(\sum_{i=1}^{n}(S_i^{(1)})^p)^{1-1/p}} \\[2mm] \quad + \dfrac{3(1-p)[\sum_{i=1}^{n}(S_i^{(1)})^{p-1}S_i^{(2)}]^2}{4(\sum_{i=1}^{n}(S_i^{(1)})^p)^{2-1/p}}, \quad \left(\sum_{i=1}^{n}(S_i^{(0)})^p\right)^{1/p} = 0, \quad \left(\sum_{i=1}^{n}(S_i^{(1)})^p\right)^{1/p} \neq 0 \\[4mm] \dfrac{\sum_{i=1}^{n}(S_i^{(2)})^{p-1}S_i^{(3)}}{(\sum_{i=1}^{n}(S_i^{(2)})^p)^{1-1/p}}, \quad \left(\sum_{i=1}^{n}(S_i^{(0)})^p\right)^{1/p} = 0, \quad \left(\sum_{i=1}^{n}(S_i^{(1)})^p\right)^{1/p} = 0, \quad \left(\sum_{i=1}^{n}(S_i^{(2)})^p\right)^{1/p} \neq 0 \\[4mm] \left(\sum_{i=1}^{n}(S_i^{(3)})^p\right)^{1/p}, \quad \left(\sum_{i=1}^{n}(S_i^{(0)})^p\right)^{1/p} = 0, \quad \left(\sum_{i=1}^{n}(S_i^{(1)})^p\right)^{1/p} = 0, \quad \left(\sum_{i=1}^{n}(S_i^{(2)})^p\right)^{1/p} = 0 \end{cases} \tag{61}$$

Then, we obtain

Theorem 9 (1≤p<∞, general case)

Let $s : \mathbb{R}_0^+ \to \mathbb{C}^n$ be an n-dimensional vector function that is m = 3 times continuously differentiable, and let $t_0 \in \mathbb{R}_0^+$.

Then,

$$\|s(t_0)\|_p = S^{(0, p)},$$

(62)

$$D_+^1 \|s(t_0)\|_p = S^{(1, p)},$$

(63)

$$D_+^2 \|s(t_0)\|_p = S^{(2, p)},$$

(64)

$$D_+^3 \|s(t_0)\|_p = S^{(3, p)},$$

(65)

where $S^{(0, p)}$-$S^{(3, p)}$ are defined by (58)–(61).

Proof
For sufficiently small $\Delta t_0 > 0$, we obtain the expansion

$$\|s(t)\|_p = S^{(0, p)} + S^{(1, p)}(t - t_0) + S^{(2, p)}\frac{(t - t_0)^2}{2!} + S^{(3, p)}\frac{(t - t_0)^3}{3!} + r(t),$$

(66)

where $r(t) = o((t-t_0)^m)$ for $t \in [t_0, t_0 + \Delta t_0]$.

Remark
It is left to the reader to show that in the case $p=1$ resp. $p=2$ we get back the results of Theorem 7 resp. Theorem 8.

Remark
There are some special cases, namely the case $p = \infty$ and the cases $p \in 2\mathbb{N}$. For, the case $p = \infty$ needs an extra treatment, and the cases $p \in 2\mathbb{N}$ are the only ones where the norms $\|x\|_p$, $\|y\|_p$, and $\|z\|_p$ are used

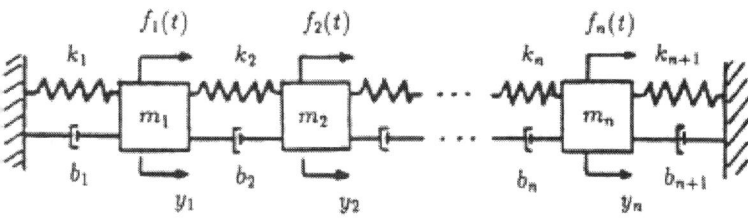

Figure 1: Multi-mass vibration model.

to distinguish the various cases in the definition of the right derivatives, and not the components x_i, y_i, and z_i. For this, see e.g. the case p=2 and compare the definition of $S^{(0)}-S^{(3)}$ according to (49)-(52) and the definition $S_i^{(0)} - S_i^{(3)}$ according to (25)-(28).

APPLICATIONS

In this section, we apply the obtained results

- to a vibration problem, first without excitation, and then with sinoidal force excitation where the amplitude is constant or linearly increasing; in the vibration problem, the first right derivative of the norm of a vector function is used, and the best upper bounds in certain classes of bounds are determined;

- To the discussion of a function $t \mapsto \|x(t)\|_2$, where the second and third right derivatives $D_+^2\|x(t)\|_2$ and $D_+^3\|x(t)\|_2$ are applied.

Upper Bounds for a Multi-Mass Vibration Problem
The Vibration Model and its Mathematical Description
Consider the multi-mass vibration model in Fig. 1.

The associated initial-value problem is given by

$$M\ddot{y} + B\dot{y} + Ky = f(t), \quad y(0) = y_0, \quad \dot{y}(0) = \dot{y}_0, \tag{67}$$

where $y = [y_1,\ldots,y_n]^\mathsf{T}$ and

$$M = \begin{bmatrix} m_1 & & & & \\ & m_2 & & & \\ & & m_3 & & \\ & & & \ddots & \\ & & & & m_n \end{bmatrix}, \tag{68}$$

$$B = \begin{bmatrix} b_1 + b_2 & -b_2 & & & & \\ -b_2 & b_2 + b_3 & -b_3 & & & \\ & -b_3 & b_3 + b_4 & -b_4 & & \\ & & \ddots & \ddots & \ddots & \\ & & & -b_{n-1} & b_{n-1} + b_n & -b_n \\ & & & & -b_n & b_n + b_{n+1} \end{bmatrix} \tag{69}$$

$$K = \begin{bmatrix} k_1 + k_2 & -k_2 & & & & \\ -k_2 & k_2 + k_3 & -k_3 & & & \\ & -k_3 & k_3 + k_4 & -k_4 & & \\ & & \ddots & \ddots & \ddots & \\ & & & -k_{n-1} & k_{n-1} + k_n & -k_n \\ & & & & -k_n & k_n + k_{n+1} \end{bmatrix} \tag{70}$$

further, the following right-hand sides are used:

$f(t) = 0$ (a) (free vibration)

$f(t) = f_0 \sin \omega t$ (b) (sinoidal force excitation with constant amplitude)

$f(t) = f_0 t \sin \omega t$ (c) (sinoidal force excitation with linearly increasing amplitude).

Using the state-space description, we obtain

$$\dot{x}(t) = Ax(t) + g(t), \quad x(0) = x_0, \tag{71}$$

with $x = [y^T, z^T]^T$, $z = \dot{y}$, where the system matrix A has the form

$$A = \left[\begin{array}{c|c} 0 & E \\ \hline -M^{-1}K & -M^{-1}B \end{array} \right]$$

(72)

and where the right-hand side is given by

$$g(t) = \left[\begin{array}{c} 0 \\ M^{-1}f(t) \end{array} \right].$$

(73)

Let

$$g_0 = \left[\begin{array}{c} 0 \\ M^{-1}f_0 \end{array} \right].$$

(74)

Then,

$$g(t) = 0 \qquad \text{in case (a)}$$
$$g(t) = g_0 \sin \omega t \qquad \text{in case (b)}$$
$$g(t) = g_0 t \sin \omega t \qquad \text{in case (c)}$$

As of now, we specify the values as

$$m_j = 1, \quad j = 1, \dots, n$$
$$k_j = 1, \quad j = 1, \dots, n$$

and

$$b_j = \begin{cases} 1/2, & j \text{ even} \\ 1/4, & j \text{ odd}. \end{cases}$$

Then,

M = E,

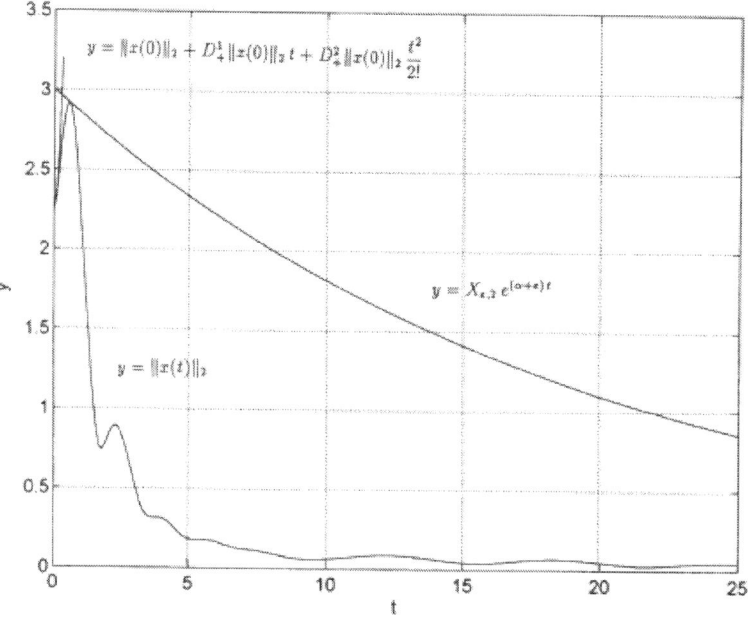

Figure 2: $y=\|x(t)\|_2$ for $f(t)=0$ and upper bounds; IC (I).

$$
B = \begin{bmatrix}
\frac{3}{4} & -\frac{1}{2} & & & & \\
-\frac{1}{2} & \frac{3}{4} & -\frac{1}{4} & & & \\
& -\frac{1}{4} & \frac{3}{4} & -\frac{1}{2} & & \\
& & \ddots & \ddots & \ddots & \\
& & & -\frac{1}{4} & \frac{3}{4} & -\frac{1}{2} \\
& & & & -\frac{1}{2} & \frac{3}{4}
\end{bmatrix}
$$

(if n is even),

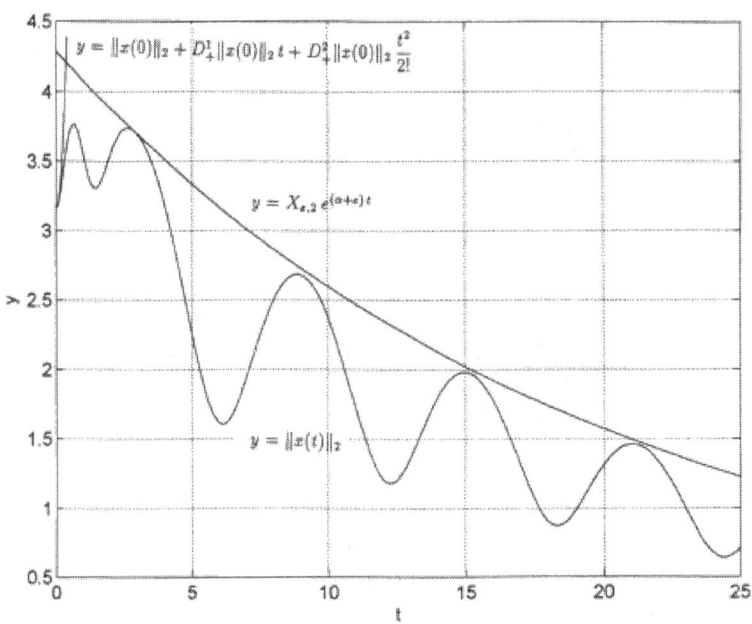

Figure 3: $y=\|x(t)\|_2$ for $f(t)=0$ and upper bounds; IC (II).

$$K = \begin{bmatrix} 2 & -1 & & & & \\ -1 & 2 & -1 & & & \\ & -1 & 2 & -1 & & \\ & & \ddots & \ddots & \ddots & \\ & & & -1 & 2 & -1 \\ & & & & -1 & 2 \end{bmatrix}.$$

Further, let

$$f_0 = [0, \dots, 0; f_{0,n}]^T,$$

where

Differential Calculus for P-Norms of Complex-Valued Vector

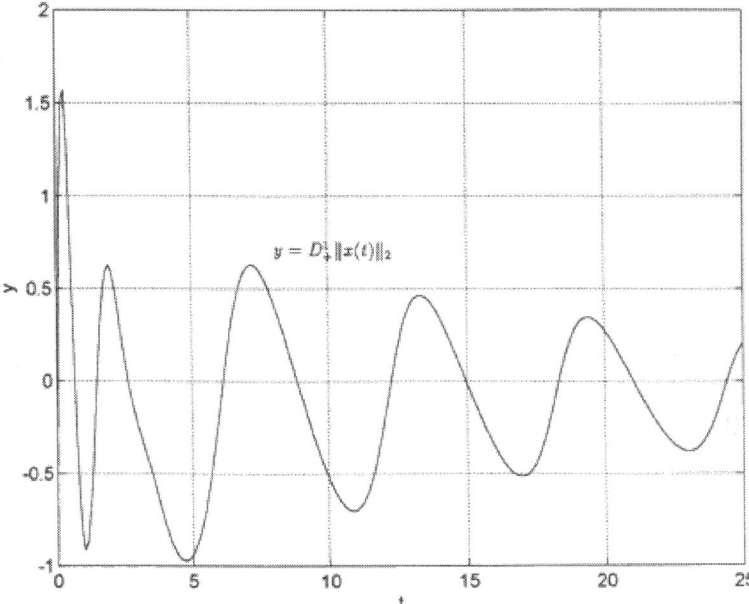

Figure 4: $y=D^1+\|x(t)\|_2$ for $f(t)=0$; IC (II).

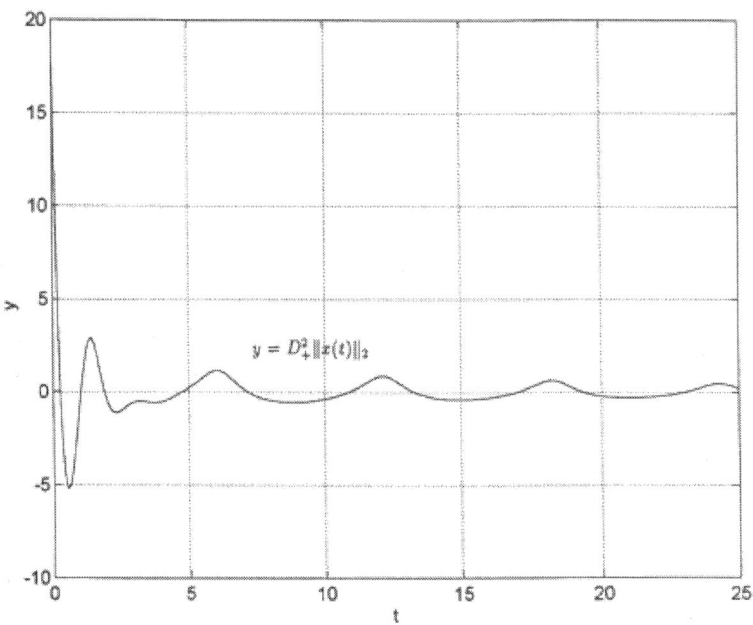

Figure 5: $y=D^2+\|x(t)\|_2$ for $f(t)=0$; IC (II).

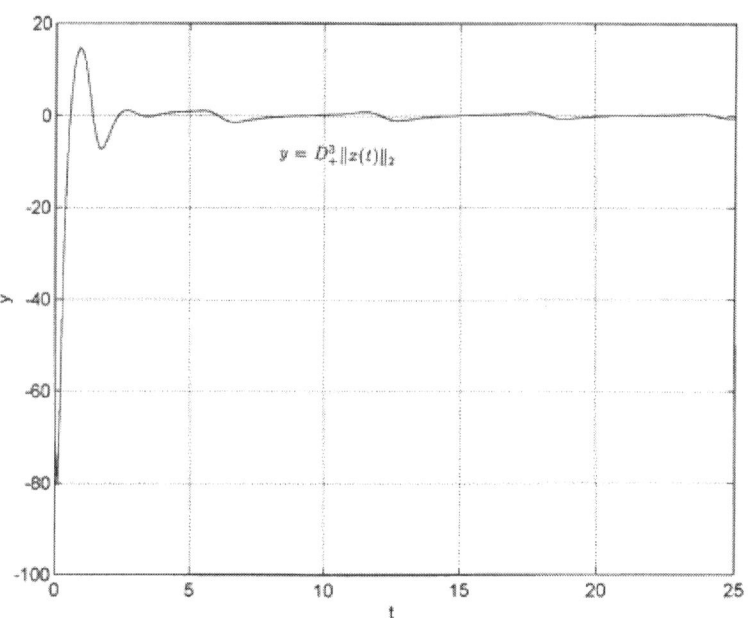

Figure 6: $y = D^3 + \|x(t)\|_2$ for $f(t)=0$; IC (II).

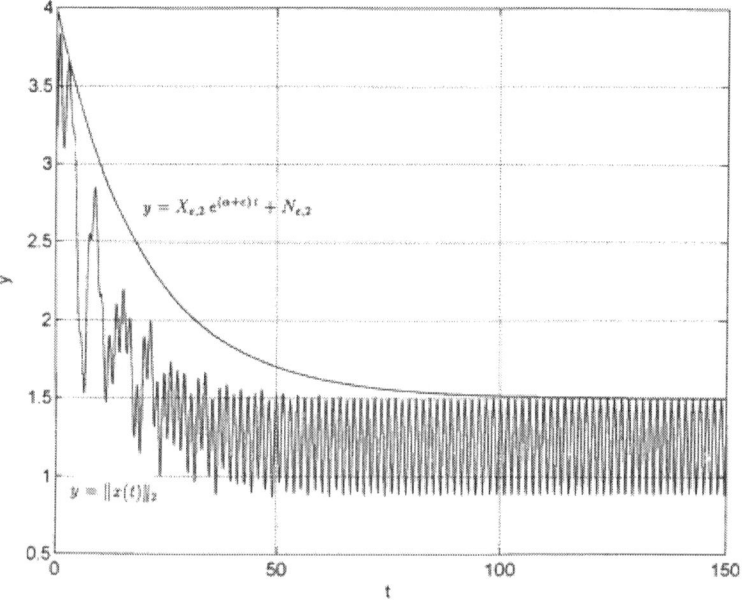

Figure 7: $y = \|x(t)\|_2$ for $f(t) = f_0 \sin \omega t$ and upper bound; IC (II).

Differential Calculus for P-Norms of Complex-Valued Vector

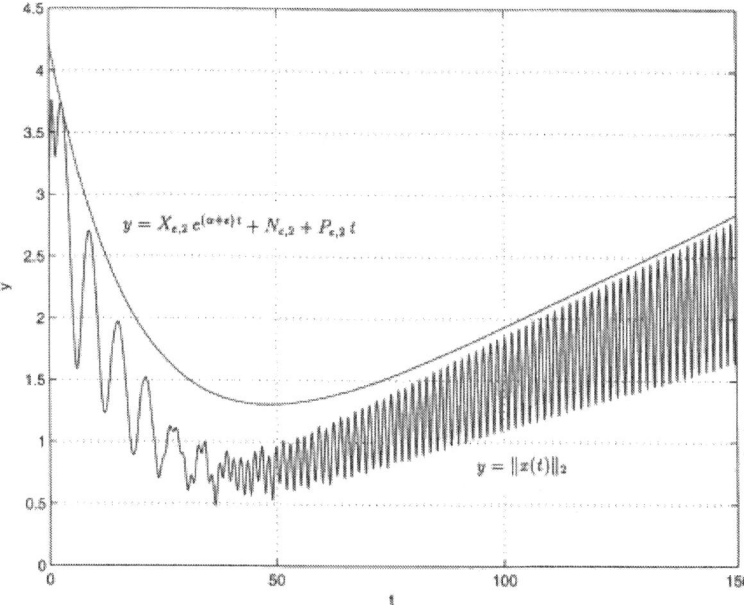

Figure 8: $y=\|x(t)\|_2$ for $f(t) = f_0 t \sin \omega t$ and upper bound; IC (II).

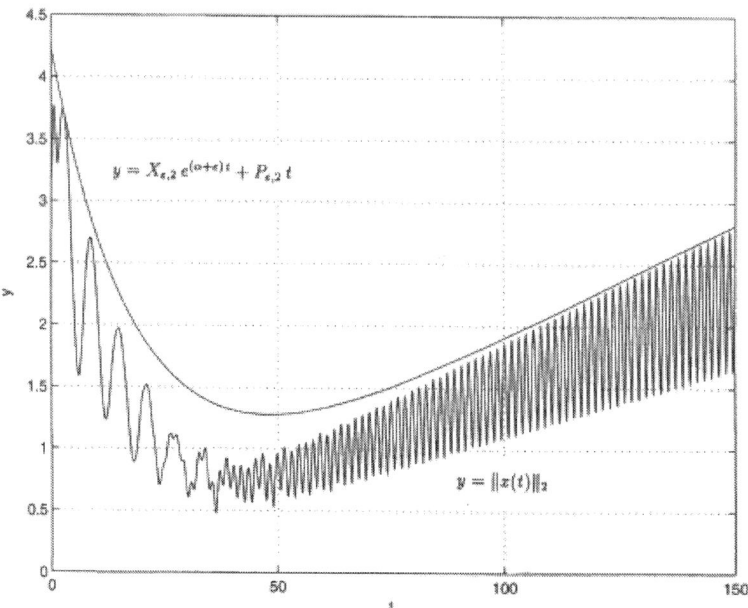

Figure 9: $y=\|x(t)\|_2$ for $f(t) = f_0 t \sin \omega t$ and upper bound; IC (II).

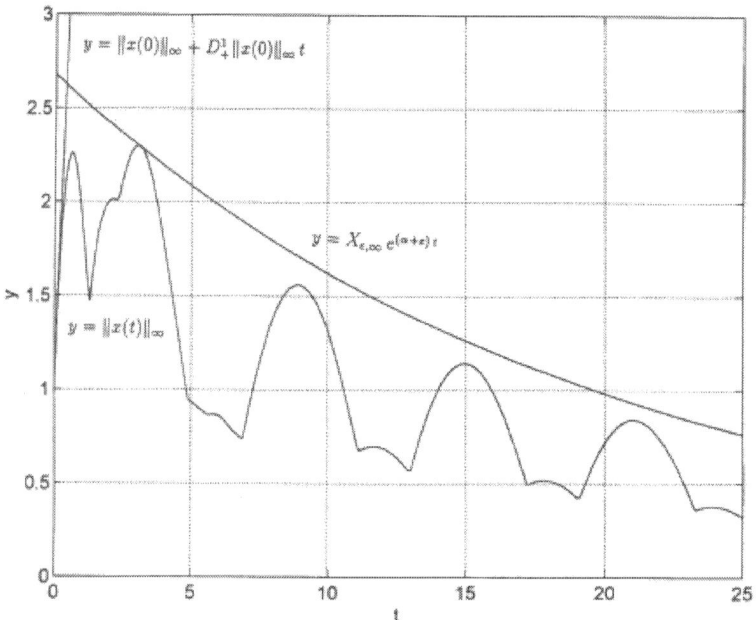

Figure 10: y=||x(t)||∞ for f(t)=0 and upper bounds; IC (II).

$$
f_{0,n} = \begin{cases} 0, & \text{in case (a)} \\ 2.0 & \text{in case (b)} \\ 0.025 & \text{in case (c).} \end{cases}
$$

We choose n=5 in this paper so that the state-space vector x(t) has the dimension m = 2n = 10.

Remark

In 1, 2 and 3, the vectors have the dimension n. The state vector x(t) has the dimension m=2n=10. But, later on in the 4.1.2 and 4.1.3, we rename the dimension of x(t) to n (instead of m=2n) where n=10. A similar practice has already been employed in the paper [11] without explicit mentioning.

The initial condition for y(t) is chosen as

$$y_0 = [-1, 1, -1, 1, -1]^T,$$

and for $\dot{y}(t)$, we consider the following two cases (I) and (II), where initial condition (I) (for short: IC (I)) is used in Fig. 2 and initial condition (II) in Fig. 3, Fig. 4, Fig. 5, Fig. 6, Fig. 7, Fig. 8, Fig. 9 and Fig. 10

$$\dot{y}_0 = [0, 0, 0, 0, 0]^T, \quad \text{in case (I)}$$

and

$$\dot{y}_0 = [-1, -1, -1, -1, -1]^T, \quad \text{in case (II)}.$$

Finally, for the excitation frequency, we choose

$$\omega = 2.0.$$

Calculations in the Norm $\|\cdot\|_2$

Case (a): free vibration: $f(t)=0$.

Solution to the problem: As is well-known, the solution to the problem $\dot{x}(t) = Ax(t), x(0) = x_0$ is given by

$$x(t) = \Phi(t)x_0 = e^{At}x_0, \tag{75}$$

where $\Phi(t)=e^{At}$ is called the fundamental matrix or evolution. We remind that the initial condition (I) is used in Fig. 2 and the initial condition (II) in Fig. 3, Fig. 4, Fig. 5, Fig. 6, Fig. 7, Fig. 8, Fig. 9 and Fig. 10.

Best upper bound on $\|x(t)\|_2$ "near the origin": In Fig. 2 and Fig. 3, $\|x(t)\|_2$ is plotted along with upper bounds; in Fig. 4, Fig. 5 and Fig. 6, $D^1 + \|x(t)\|_2$, $D^2 + \|x(t)\|_2$, and $D^3 + \|x(t)\|_2$ are shown.

Both in Fig. 2 and Fig. 3, the upper bounds in the interval $[0, t_2^*]$ "near the origin" are given by

$$\|x(t)\|_2 \leqslant \|x(0)\|_2 + D_+^1 \|x(0)\|_2 t + D_+^2 \|x(0)\|_2 \frac{t^2}{2!}, \tag{76}$$

where in Fig. 2,

$$\|x(0)\|_2 \doteq 2.2361$$

$$D_+^1 \|x(0)\|_2 = 0$$

$$D_+^2 \|x(0)\|_2 \doteq 21.4663$$

and in Fig. 3,

$$\|x(0)\|_2 \doteq 3.1623$$

$$D_+^1 \|x(0)\|_2 \doteq -0.5534$$

$$D_+^2 \|x(0)\|_2 \doteq 18.1258.$$

These upper bounds are the best possible ones in the class of polynomials $y = \sum_{k=0}^{N} D^1 + \| x(0) \|_2 \, t^k / k!$ with $N \in \mathbb{N}$. For Fig. 3, this is seen from the Fig. 3, Fig. 4 and Fig. 5 for Fig. 2, this can be concluded from the figures for $D^1 + \|x(t)\|_2$ and $D^2 + \|x(t)\|_2$, which are not shown here for the sake of brevity.

Best upper bound on $\|x(t)\|_2$ "far from the origin" (that is, in the adjacent interval $[t_2^*, \infty]$). Again, for every $\varepsilon > 0$ there exists a constant $M_{\varepsilon,2} > 0$ such that

$$\|\Phi(t)\|_2 \leqslant M_{\varepsilon,2} \, e^{(\alpha+\varepsilon)t}, \quad t \geqslant 0,$$

where α is the spectral abscissa of the matrix A. As in [11], in the programmes we have chosen the machine precision ε =eps=$2^{-52} \doteq 2.2204 \times 10^{-16}$ of MATLAB. From $x(t) = \Phi(t)x_0 = e^{At}x_0$, we obtain

$$\|x(t)\|_2 \leqslant X_{\varepsilon,2} \, e^{(\alpha+\varepsilon)t} \tag{77}$$

with

$$X_{\varepsilon,2} = M_{\varepsilon,2} \|x_0\|_2 \doteq 4.4026 \quad \text{resp.} \doteq 6.2262 \tag{78}$$

for IC (I) resp. IC (II) since $M_{\varepsilon,2} \doteq 1.9689$ (see [11]) and $\|x_0\|_2 = \sqrt{5}$ if IC (I) resp. $\|x_0\|_2 = \sqrt{10}$ if IC (II).

These constants $X_{\varepsilon,2}$ are not optimal, however. To determine the minimal $X_{\varepsilon,2}$, let $\|\cdot\|$ be any vector norm, for which $D^1 + \|x(t)\|$ can be calculated. To obtain the minimal $X_{\varepsilon,2}$ such that $\|x(t)\| \leq X_\varepsilon e^{(\alpha+\varepsilon)t}, t \geq 0$, we proceed similarly as in [11] and seek a place t_c where the function

$$t \mapsto x_\varepsilon(t) := X_\varepsilon e^{(\alpha+\varepsilon)t}, \quad t \geq 0 \tag{79}$$

meets the function $t \mapsto \|x(t)\|$. Thus,

$$\|x(t_c)\| \stackrel{!}{=} x_\varepsilon(t_c) \tag{80}$$

and

$$D^1_+ \|x(t_c)\| \stackrel{!}{=} x'_\varepsilon(t_c) = (\alpha + \varepsilon) x_\varepsilon(t_c). \tag{81}$$

After elimination of $x\varepsilon(tc)$, we obtain

$$D^1_+ \|x(t_c)\| \stackrel{!}{=} (\alpha + \varepsilon) \|x(t_c)\|, \tag{82}$$

which is a single nonlinear equation in the single unknown t_c.

After t_c has been determined numerically from (82), we get $X\varepsilon$ from the relation

$$X_\varepsilon = \|x(t_c)\| e^{-(\alpha+\varepsilon)t_c}. \tag{83}$$

This is now applied to the case $\|x(\cdot)\| = \|x(\cdot)\|_2$. In Fig. 2 (for initial condition (I)), the point of contact $t_{c,2}$ is near $t=1.0$. We obtain

$\alpha \doteq -0.0502,$

and

$$t_{c,2} \doteq 0.5652,$$

$$X_{\varepsilon,2} \doteq 3.0047,$$

$$t_2^* \doteq 0.2607,$$

where t_2^* is the abscissa of the intersection point between the two upper bounds. In Fig. 3 (for initial condition (II)), the point of contact $t_{c,2}$ is near $t = 3.0$. We obtain

$$t_{c,2} \doteq 2.8795,$$

$$X_{\varepsilon,2} \doteq 4.2918,$$

$$t_2^* \doteq 0.3723.$$

Both values of $X_{\varepsilon,2}$ are smaller than those in (78), respectively.

Case (b): Sinoidal force excitation with constant amplitude: $f(t) = f_0 \sin \omega t$.

Solution to the problem: The solution to the problem

$$\dot{x}(t) = A x(t) + g_0 \sin \omega t, \quad x(0) = x_0 \tag{84}$$

is given by

$$x(t) = e^{At} c_0 + a_0 \sin \omega t + b_0 \cos \omega t \tag{85}$$

with

$$c_0 = x_0 + \omega(A^2 + \omega^2 E)^{-1} g_0 \tag{86}$$

$$a_0 = -A(A^2 + \omega^2 E)^{-1} g_0 \tag{87}$$

$$b_0 = -\omega(A^2 + \omega^2 E)^{-1} g_0. \tag{88}$$

Best upper bound on $\|x(t)\|_2$ "far from the origin" (we restrict ourselves to this case). Starting with (85), we obtain

$$\|x(t)\|_2 \leqslant \|e^{A\,t}\|_2\|c_0\|_2 + \|a_0 \sin \omega t + b_0 \cos \omega t\|_2$$

$$\leqslant X_{\varepsilon,2}\, e^{(\alpha+\varepsilon)t} + N_{\varepsilon,2} \tag{89}$$

with

$$X_{\varepsilon,2} = M_{\varepsilon,2}\|c_0\|_2 \doteq 1.9689 \times 2.8563 \doteq 5.6238 \tag{90}$$

and

$$N_{\varepsilon,2} = \max_{t\geqslant 0} \|a_0 \sin \omega t + b_0 \cos \omega t\|_2. \tag{91}$$

The argument ωt_{max} such that

$$N_{\varepsilon,2} = \max_{t\geqslant 0} \|a_0 \sin \omega t + b_0 \cos \omega t\|_2$$

$$= \|a_0 \sin(\omega t_{max}) + b_0 \cos(\omega t_{max})\|_2 \tag{92}$$

is given by

$$\omega t_{max} = \frac{1}{2} \arctan\left(\frac{2(a,b)}{\|b\|_2^2 - \|a\|_2^2}\right) \quad \text{with } a = a_0,\ b = b_0. \tag{93}$$

Thereby, $\omega t_{max} \doteq -0.7890$ (or $\omega t_{max} = -0.7890 + 2\pi \doteq 5.4942$ if one wants a positive value for t) and thus

$$N_{\varepsilon,2} \doteq 1.5025. \tag{94}$$

The value of $X_{\varepsilon,2}$ in (90) is not optimal, however. In order to determine the minimal value for $X_{\varepsilon,2}$ in (89), we proceed similarly as above in the case (a). So, let $\|\cdot\|$ be any vector norm, for which $D^1 + \|\cdot\|$ can be determined. Then, we seek a place t_c at which the function

$$t \mapsto x_\varepsilon(t) := X_\varepsilon\, e^{(\alpha+\varepsilon)t} + N_\varepsilon, \quad t \geqslant 0 \tag{95}$$

meets the function x(t) in (85). Thus,

$$\|x(t_c)\| \overset{!}{=} x_\varepsilon(t_c) = X_\varepsilon\, e^{(\alpha+\varepsilon)t_c} + N_\varepsilon \tag{96}$$

and

$$D^1_+ \|x(t_c)\| \overset{!}{=} x'_\varepsilon(t_c) = (\alpha + \varepsilon)X_\varepsilon\, e^{(\alpha+\varepsilon)t_c}. \tag{97}$$

These equations lead to

$$D^1_+ \|x(t_c)\| \overset{!}{=} (\alpha + \varepsilon)(\|x(t_c)\| - N_\varepsilon), \tag{98}$$

which is a single nonlinear equation in the single unknown t_c.

After t_c has been determined numerically from (98), we get Xε from the relation

$$X_\varepsilon = (\|x(t_c)\| - N_\varepsilon)e^{-(\alpha+\varepsilon)t_c}. \tag{99}$$

This is now applied to the case $\|x(\cdot)\| = \|x(\cdot)\|_2$. We obtain the following results

$$N_{\varepsilon,2} \doteq 1.5025,$$

and

$$t_{c,2} \doteq 2.7445,$$

$$X_{\varepsilon,2} \doteq 2.4825,$$

and thus

$$x_{\varepsilon,2}(t = 0) = X_{\varepsilon,2} + N_{\varepsilon,2} = 3.9850.$$

In Fig. 7, $\|x(t)\|_2$ for the case (b) is plotted along with the best upper bound $x\varepsilon_{,2}(t)$ with a stepsize $\Delta t = 0.1$.

Case (c): Sinoidal force excitation with linearly increasing amplitude:

$$f(t) = f_0 t \sin\omega t.$$

Solution to the problem: According to Section A.1, the solution to the problem

$$\dot{x}(t) = Ax(t) + g_0 t \sin \omega t, \quad x(0) = x_0 \tag{100}$$

is given by

$$x(t) = e^{At} c_0 + a_0 \sin \omega t + b_0 \cos \omega t$$
$$+ t(a_1 \sin \omega t + b_1 \cos \omega t) \tag{101}$$

with

$$c_0 = x_0 + 2\omega A (A^2 + \omega^2 E)^{-2} g_0, \tag{102}$$
$$a_0 = -(A^2 - \omega^2 E)(A^2 + \omega^2 E)^{-2} g_0, \tag{103}$$
$$b_0 = -2\omega A (A^2 + \omega^2 E)^{-2} g_0, \tag{104}$$
$$a_1 = -A(A^2 + \omega^2 E)^{-1} g_0, \tag{105}$$
$$b_1 = -\omega(A^2 + \omega^2 E)^{-1} g_0. \tag{106}$$

Best upper bound on $\|x(t)\|_2$ "far from the origin" (we restrict ourselves to this case). Starting with (101), we obtain

$$\|x(t)\|_2 \leqslant \|e^{At}\|_2 \|c_0\|_2 + \|a_0 \sin \omega t + b_0 \cos \omega t\|_2 + t\|a_1 \sin \omega t + b_1 \cos \omega t\|_2$$
$$\leqslant X_{\varepsilon,2} e^{(\alpha+\varepsilon)t} + N_{\varepsilon,2} + P_{\varepsilon,2} t \tag{107}$$

with

$$X_{\varepsilon,2} = M_{\varepsilon,2} \|c_0\|_2 \doteq 1.9689 \times 3.1668 \doteq 6.2351 \tag{108}$$

and

$$N_{\varepsilon,2} = \max_{t \geqslant 0} \|a_0 \sin \omega t + b_0 \cos \omega t\|_2 \tag{109}$$

$$P_{\varepsilon,2} = \max_{t \geqslant 0} \|a_1 \sin \omega t + b_1 \cos \omega t\|_2. \tag{110}$$

The argument ωt_{max} resp. $\omega t_{max}{}'$ such that

$$N_{\varepsilon,2} = \|a_0 \sin(\omega t_{max}) + b_0 \cos(\omega t_{max})\|_2 \tag{111}$$

resp.

$$P_{\varepsilon,2} = \|a_1 \sin(\omega t'_{max}) + b_1 \cos(\omega t'_{max})\|_2 \tag{112}$$

is given by (93) with $a = a_0, b = b_0$ resp. $a = a_1, b = b_1$.

So, $N\varepsilon_{,2}$ and $P\varepsilon_{,2}$ are known.

The value of $X\varepsilon_{,2}$ in (108) is not optimal, however. In order to determine the minimal value for $X\varepsilon_{,2}$, we proceed similarly as above in the case (b). So, let $\|\cdot\|$ be any vector norm, for which $D^1 + \|x(t)\|$ can be determined. Then, we seek a place t_c at which the function

$$t \mapsto x_\varepsilon(t) := X_\varepsilon e^{(\alpha+\varepsilon)t} + N_\varepsilon + P_\varepsilon t, \quad t \geqslant 0 \tag{113}$$

meets the function x(t) in (101). Thus,

$$\|x(t_c)\| \overset{!}{=} x_\varepsilon(t_c) = X_\varepsilon e^{(\alpha+\varepsilon)t_c} + N_\varepsilon + P_\varepsilon t_c \tag{114}$$

and

$$D^1_+ \|x(t_c)\| \overset{!}{=} x'_\varepsilon(t_c) = (\alpha + \varepsilon) X_\varepsilon e^{(\alpha+\varepsilon)t_c} + P_\varepsilon. \tag{115}$$

These equations lead to

$$D^1_+ \|x(t_c)\| - P_\varepsilon \overset{!}{=} (\alpha + \varepsilon)(\|x(t_c)\| - N_\varepsilon - P_\varepsilon t_c), \tag{116}$$

which is a single nonlinear equation in the single unknown t_c.

After t_c has been determined numerically from (116), we get $X\varepsilon$ from the relation

$$X_\varepsilon = (\|x(t_c)\| - N_\varepsilon - P_\varepsilon t_c)e^{-(\alpha+\varepsilon)t_c}. \tag{117}$$

This is now applied to the case $\|x(\cdot)\| = \|x(\cdot)\|_2$. We obtain the following results:

$N_{\varepsilon,2} \doteq 0.0240,$

$P_{\varepsilon,2} \doteq 0.0188,$

and

$$t_{c,2} \doteq 2.8519,$$
$$X_{\varepsilon,2} \doteq 4.2035,$$

and thus

$$x_{\varepsilon,2}(t=0) = X_{\varepsilon,2} + N_{\varepsilon,2} = 4.2275.$$

In Fig. 8, $\|x(t)\|_2$ for the case (c) is plotted along with the best upper bound $x_{\varepsilon,2}(t)$ with a stepsize $\Delta t=0.1$.

Remark

From Fig. 8, it is clear that $N_{\varepsilon,2} + P_{\varepsilon,2}t$ is not the least upper bound on the norm of the stationary part of the solution, $\|x_{st}(t)\|_2$. So, the calculated value $X_{\varepsilon,2}$ is only optimal with respect to the given values for $N_{\varepsilon,2}$ and $P_{\varepsilon,2}$. The least upper bound is obtained for $N_{\varepsilon,2}=0$ (cf. Fig. 9) since $P_{\varepsilon,2}t$ is an asymptote of $\|x_{st}(t)\|_2$. Then, $t_{c,2} \doteq 2.8539$ and $X_{\varepsilon,2}=x_{\varepsilon,2}(t=0) \doteq 4.2311$.

Calculations in the norm $\|\cdot\|_\infty$

For the sake of brevity, we restrict ourselves to the plots of $\|x(t)\|_\infty$ in the cases (a) and (c) along with their best upper bounds $x_\varepsilon(t)$, which are computed similarly as for $p=2$.

Case (a): Free vibration: $f(t)=0$.

Best upper bound on $\|x(t)\|_\infty$ "near the origin": In Fig. 10, the upper bound in the interval $[0,t_2^*]$ "near the origin" is given by

$$\|x(t)\|_\infty \leqslant \|x(0)\|_\infty + D_+^1\|x(0)\|_\infty t$$

where

$$\|x(0)\|_\infty = 1,$$
$$D_+^1\|x(0)\|_\infty = 4.$$

This upper bound is the best one in the class of polynomials $y = \sum_{k=0}^{N} D_+^k \| x(0) \|_\infty t^k / k!$ with $N \in \mathbb{N}$. This is justified by Fig. 10 and the plots for $D^1 + \| x(t) \|_\infty$ and $D^2 + \| x(t) \|_\infty$, which are not shown here for the sake of brevity. (These plots are similar to those of $D^1 + \| \Phi(t) \|_\infty$ and $D^2 + \| \Phi(t) \|_\infty$ in [11]; but there is a difference: $D^2 + \| x(t) \|_\infty$ is positive in a small region, as opposed to $D^2 + \| \Phi(t) \|_\infty$, which is negative throughout the considered interval.)

Best upper bound on $\| x(t) \|_\infty$ "far from the origin" (that is, in the adjacent interval $[t^*_\infty, \infty]$). Here, starting with $x(t) = e^{At} x_0$, we obtain

$$\| x(t) \|_\infty \leqslant X_{\varepsilon, \infty} e^{(\alpha + \varepsilon)t} \tag{118}$$

where

$$X_{\varepsilon, \infty} = M_{\varepsilon, \infty} \| x_0 \|_\infty \doteq 3.1148 \times 1 \doteq 3.1148 \tag{119}$$

(for $M\varepsilon_{,\infty} \doteq 3.1148$, see [11]).

This constant $X\varepsilon_{,\infty}$ is not optimal, however. The minimal $X\varepsilon_{,\infty}$ is computed as for $p=2$; we obtain the following numerical values:

$t_{c,\infty} \doteq 3.1291,$

$X_{\varepsilon, \infty} \doteq 2.6826,$

$t^*_\infty \doteq 0.4071.$

So, we get an improvement for $X\varepsilon_{,\infty}$.

Case (c): Sinoidal force excitation with linearly increasing amplitude: $f(t) = f_0 t \sin \omega t$.

Best upper bound on $\| x(t) \|_\infty$ "far from the origin" (we restrict ourselves to this bound): Starting with (101), we obtain

$$\| x(t) \|_\infty \leqslant X_{\varepsilon, \infty} e^{(\alpha + \varepsilon)t} + N_{\varepsilon, \infty} + P_{\varepsilon, \infty} t \quad t \geqslant 0 \tag{120}$$

with

$$X_{\varepsilon, \infty} = M_{\varepsilon, \infty} \| c_0 \|_\infty \doteq 3.1148 \times 1.0193 \doteq 3.1749 \tag{121}$$

and

$$N_{\varepsilon,\infty} = \max_{t \geq 0} \|a_0 \sin \omega t + b_0 \cos \omega t\|_\infty, \tag{122}$$

$$P_{\varepsilon,\infty} = \max_{t \geq 0} \|a_1 \sin \omega t + b_1 \cos \omega t\|_\infty. \tag{123}$$

For vectors $a, b \in \mathbb{R}^n$, one has

$$\max_{t \geq 0} \|a \sin \omega t + b \cos \omega t\|_\infty = \max_{j=1,\ldots,n} |a_j \sin \omega t_j + b_j \cos \omega t_j| \tag{124}$$

with

$$\omega t_j = \frac{1}{2} \arctan\left(\frac{2a_j b_j}{b_j^2 - a_j^2}\right), \quad j = 1,\ldots,n \tag{125}$$

if $b_j^2 \neq a_j^2, j = 1,\ldots,n$. To determine $N_{\varepsilon,\infty}$ resp. $P_{\varepsilon,\infty}$, the relations (124), (125) are applied with $a = a_0, b = b_0$ resp. $a = a_1, b = b_1$.

So, $N_{\varepsilon,\infty}$ and $P_{\varepsilon,\infty}$ are known.

The minimal value of $X_{\varepsilon,\infty}$ is computed as for p=2; we obtain the numerical results:

$$N_{\varepsilon,\infty} \doteq 0.0194,$$
$$P_{\varepsilon,\infty} \doteq 0.0181,$$

and hereby

$$t_{c,\infty} \doteq 3.1118,$$
$$X_{\varepsilon,\infty} \doteq 2.5902,$$

and thus

$$x_{\varepsilon,\infty}(t = 0) = X_{\varepsilon,\infty} + N_{\varepsilon,\infty} = 2.6096.$$

In Fig. 11, $\|x(t)\|_\infty$ for the case (c) is plotted along with the best upper bound $x_{\varepsilon,\infty}(t)$ with a stepsize $\Delta t = 0.1$.

Remark

A similar remark as for p=2 holds with regard to the upper bound $N\varepsilon_{\infty}+P\varepsilon_{,\infty}t$. For $N\varepsilon_{,\infty}=0$ (cf. Fig. 12), we obtain an asymptote of the stationary part $\|x_{st}(t)\|_{\infty}$ as well as $t_{c,\infty} \doteq 3.1126$ and $X\varepsilon_{,\infty}=x\varepsilon_{,\infty}(t=0) \doteq 2.6129$.

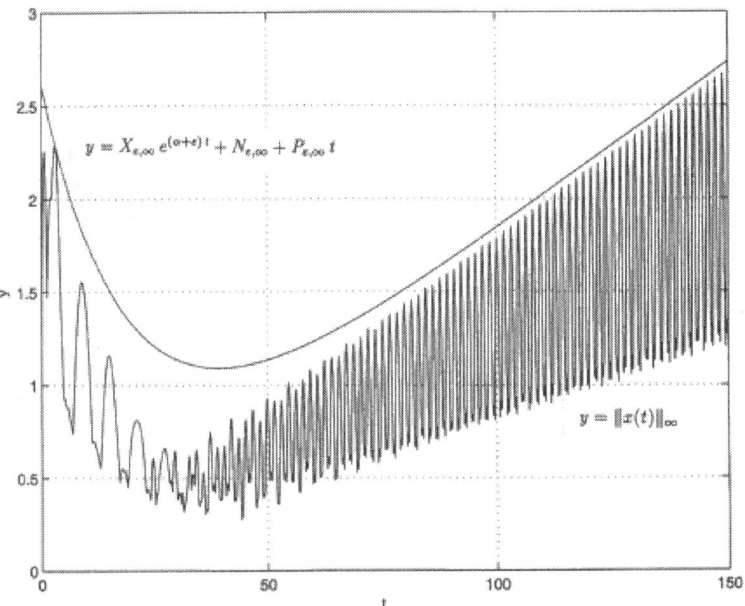

Figure 11: $y=\|x(t)\|_{\infty}$ for $f(t) = f_0 t \sin \omega t$ and upper bound; IC (II).

Discussion of a Function $t \mapsto \|x(t)\|p$

We determine the inflexion points of the function $t \mapsto \|x(t)\|_2$ for initial condition (II) (cf. Fig. 3). For this, we need the second and third right derivatives $D^2+\|x(t)\|_2$ and $D^3+\|x(t)\|_2$. We consider the range $0 \leq t \leq 25$. The conditions $D^2+\|x(t_{ip})\|_2=0$ and $D^3+\|x(t_{ip})\|_2 \neq 0$ deliver the inflexion points. We then obtain Table 1.

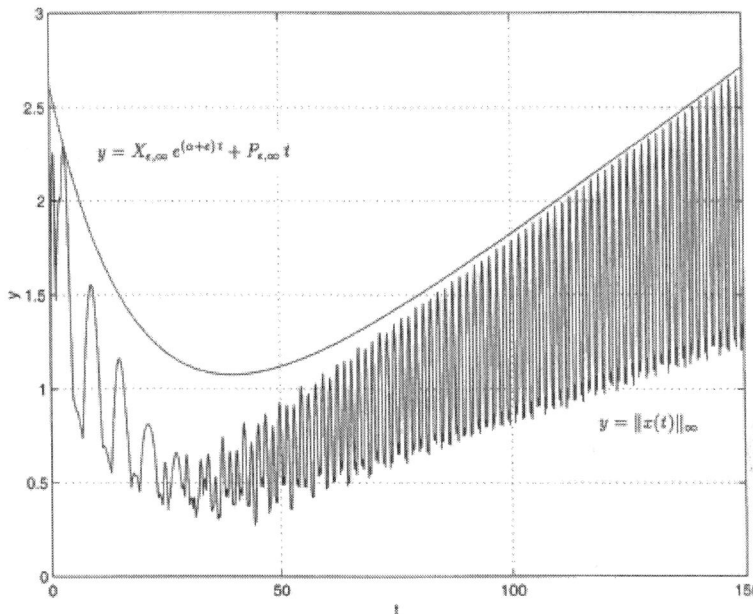

Figure 12: $y=\|x(t)\|_\infty$ for $f(t)=f_0 t\sin\omega t$ and upper bound; IC (II).

Table 1: Abscissae of inflexion points and pertinent third derivatives

xip	D3+‖x(tip)‖2
0.2577	−45.6162
1.0122	14.0181
1.8865	−5.7192
4.7228	0.8455
7.2042	−0.8742
10.8503	0.5897
13.2842	−0.6341
16.9342	0.4339
19.3685	−0.4671
23.0179	0.3196

CONCLUDING REMARKS

The differential calculus for p-norms of complex-valued vector functions has turned out to be a powerful and elegant method for obtaining optimal results in the area of upper bounds on the norm of a vector function. These results with the minimal constants in the considered estimates cannot be achieved by the methods used so far, in general. So far, similar results could only be obtained for one-dimensional problems (cf. [18, p. 28, Fig. 2.2-3 or p. 30. Fig. 2.4-1]).

We remark that [10, Lemma 2.1] is needed in the case of vector functions only for p=∞, whereas in the case of the fundamental matrix in [11], it is needed for all $p \in \{1, 2, \infty\}$ simply because the maximum of n numbers is formed there for all these p. Further, we mention in passing that, in the cited Lemma 2.1 of [10], the functions need only be continuous so that it is applicable here.

Since $\max\{||y(t)||, ||\dot{y}(t)||\} \leq ||x(t)||$ in Section 4, one can also compute the best upper bounds on the displacement ‖y(t)‖ and on the velocity $||\dot{y}(t)||$.

Starting point of the author's investigation on this subject has been the work of Coppel, Dahlquist, Desoer/Haneda and Lozinskiĭ, whose papers [2], [3], [4] and [12] are therefore added to the list of references even though these papers are not directly used here. Also, [5], [7], [8], [14], [15], [16] and [17] contain material related to the logarithmic derivative.

To obtain examples for optimal upper bounds on the solution of differential equations of a more general type (say, for a differential equation with time-dependent system matrix or for a nonlinear differential equation), in every case one has to proceed, in principle, as follows:

- Derive an upper bound on the solution that defines the class of the bound; this upper bound has one or more constants.

- Determine the best constant(s): If the solution is calculated by a step-by-step method on a grid, the differential calculus cannot be applied, but the constant(s) can be determined in a rather simple way. If the solution (supposed to be sufficiently smooth) and the upper bound are found on a whole interval, the best upper bound may be determined by the differential calculus just as in [11] and in this paper, at least in principle.

- More cannot be said at the moment because further research is needed.

The issues of step-by-step methods for the solution of the differential equation and of time-dependent matrices A(t) will be addressed in subsequent papers.

ACKNOWLEDGEMENTS

The author would like to thank the referee for his/her comments, especially for the proposal to include the general case 1<p<∞.

APPENDIX A

A.1: Determination of the Solution in the Case (C): $f(t) = f_0 t \sin \omega t$

The state-space description of the initial-value problem is determined by (71), that is,

$$\dot{x}(t) = Ax(t) + g(t), \quad x(0) = x_0 \tag{A.1}$$

with

$$g(t) = g_0 t \sin \omega t. \tag{A.2}$$

According to [13, p. 75], its solution is given by the formula

$$x(t) = \Phi(t)x_0 + \int_0^t \Phi(t - \tau)g(\tau)\,d\tau \tag{A.3}$$

or

$$x(t) = \Phi(t)x_0 + \int_0^t \Phi(t - \tau)\tau \sin \omega \tau\,d\tau g_0, \tag{A.4}$$

that is,

$$x(t) = e^{At}x_0 + e^{At} \int_0^t \tau e^{-A\tau} \sin \omega\tau \, d\tau g_0.$$

(A.5)

Now, according to [1, p. 327, no. 463], for numbers $a, b \in \mathbb{C}$ and $x \in \mathbb{R}$,

$$\int x e^{ax} \sin bx \, dx = \frac{x e^{ax}}{a^2 + b^2}(a \sin bx - b \cos bx)$$

$$- \frac{e^{ax}}{(a^2 + b^2)^2}[(a^2 - b^2)\sin bx - 2ab\cos bx].$$

(A.6)

Correspondingly, a similar formula holds for integrals involving a matrix A, namely,

$$\int \tau e^{-A\tau} \sin \omega\tau \, d\tau = [A^2 + \omega^2 E]^{-1} \tau e^{-A\tau}(-A \sin \omega\tau - E\omega \cos \omega\tau)$$

$$- [A^2 + \omega^2 E]^{-2} e^{-A\tau}[(A^2 - \omega^2 E)\sin \omega\tau + 2A\omega \cos \omega\tau]$$

(A.7)

(which may also be derived rigorously using integration by parts). This entails

$$\int_0^t \tau e^{-A\tau} \sin \omega\tau \, d\tau$$

$$= [A^2 + \omega^2 E]^{-1} t e^{-At}[-A \sin \omega t - E\omega \cos \omega t]$$

$$- [A^2 + \omega^2 E]^{-2}\{e^{-At}[(A^2 - \omega^2 E)\sin \omega t + 2A\omega \cos \omega t] - 2A\omega E\}.$$

(A.8)

By substituting (A.8) into (A.5), we finally obtain

$$x(t) = e^{At}c_0 + a_0 \sin \omega t + b_0 \cos \omega t + t(a_1 \sin \omega t + b_1 \cos \omega t)$$

(A.9)

with

$$c_0 = x_0 + 2\omega A(A^2 + \omega^2 E)^{-2}g_0,$$

(A.10)

$$a_0 = -(A^2 - \omega^2 E)(A^2 + \omega^2 E)^{-2}g_0,$$

(A.11)

$$b_0 = -2\omega A(A^2 + \omega^2 E)^{-2} g_0,$$ (A.12)

$$a_1 = -A(A^2 + \omega^2 E)^{-1} g_0,$$ (A.13)

$$b_1 = -\omega(A^2 + \omega^2 E)^{-1} g_0.$$ (A.14)

We note that $(A^2+\omega^2 E)^{-1}$ exists since $-\omega^2$ is not an eigenvalue of A^2, because $B\neq 0$.

A.2 Different Representation of Dk+||s(t0)||2

In order to obtain a different representation of $D^k+\|s(t_0)\|_2$ resembling more that of $D^k+\|s(t_0)\|_\infty$, rewrite , , and . First, note that

$$\operatorname{Re}(x, y) = \operatorname{Re} \sum_{i=1}^{n} x_i \bar{y}_i = \sum_{i=1}^{n} (\operatorname{Re} x_i \operatorname{Re} y_i + \operatorname{Im} x_i \operatorname{Im} y_i)$$ (A.15)

and

$$\|x\|_2 = \sum_{i=1}^{n} (|x_i|^2)^{1/2}.$$ (A.16)

Hereby, for example, (49) and (50) take the form

$$S^{(0)} := \sum_{i=1}^{n} (|x_i|^2)^{1/2},$$ (A.17)

$$S^{(1)} := \begin{cases} \dfrac{\sum_{i=1}^{n}(\operatorname{Re} x_i \operatorname{Re} y_i + \operatorname{Im} x_i \operatorname{Im} y_i)}{\sum_{i=1}^{n}(|x_i|^2)^{1/2}}, & x \neq 0 \\ \displaystyle\sum_{i=1}^{n}(|y_i|^2)^{1/2}, & x = 0. \end{cases}$$ (A.18)

We leave it to the reader to rewrite also formulae (51) and (52).

A.3. Different Representation of Dk+||s(t0)||∞

In order to obtain a different representation of $D^k+\|s(t_0)\|_\infty$ resembling more that of $D^k+\|s(t_0)\|_2$, rewrite (25)-(28). First, note that the scalar product of the complex numbers x_i and y_i is given by $(x_i, y_i) = x_i \bar{y}_i$.

Thus, for the real part we obtain $\text{Re}(x_i, y_i) = \text{Re}\, x_i \text{Re}\, y_i + \text{Im}\, x_i \text{Im}\, y_i$ and therefore,

$$S_i^{(0)} := |x_i|,$$

(A.19)

$$S_i^{(1)} := \begin{cases} \dfrac{\text{Re}(x_i, y_i)}{|x_i|}, & x_i \neq 0, \\ |y_i|, & x_i = 0. \end{cases}$$

(A.20)

We leave it to the reader to rewrite also formulae (27) and (28).

REFERENCES

1. I. Bronstein, K. Semendjajew, Taschenbuch der Mathematik, Verlag Harri Deutsch, ZQurich und Frankfurt=M, 1967.
2. W.A. Coppel, Stability and Asymptotic Behavior of Differential Equations, D.C. Heath and Company, Boston, 1965.
3. G. Dahlquist, Stability and Error Bounds in the Numerical Integration of Ordinary Differential Equations, Transactions of the Royal Institut of Technology, Stockholm, 1959.
4. Ch.A. Desoer, H. Haneda, The measure of a matrix as a tool to analyse computer algorithms for circuit analysis, IEEE Trans. Circuit Theory 19 (5) (1972) 480–486.
5. E. Hairer, S.P. NHrset, G. Wanner, Solving Ordinary Differential Equations I, Springer, Berlin, 1993.
6. H. Heuser, Lehrbuch der Analysis, Teil 1, B. G. Teubner, Stuttgart, 1986.
7. I. Higueras, B. GarcYZa-Celayeta, Logarithmic norms for matrix pencils, SIAM J. Matrix Anal. Appl. 20 (1999) 646–666.
8. I. Higueras, B. GarcYZa-Celayeta, How close can the logarithmic norm of a matrix pencil come to the spectral abscissa?, SIAM J. Matrix Anal. Appl. 22 (2000) 472–478.
9. K. Knopp, Theory of functions Parts I and II, Dover Publications, New York, 1945 and 1947, (English Translation).
10. L. Kohaupt, Second Logarithmic Derivative of a Complex Matrix in the Chebyshev Norm, SIAM J. Matrix Anal. Appl. 21 (2) (1999) 382–389.
11. L. Kohaupt, Differential calculus for some p-norms of the fundamental matrix with applications, J. Comput. Appl. Math. 135 (2001) 1–21.
12. S.M. LozinskiUi, Error estimates for the numerical integration of ordinary differential equations, I. Izv. Vys U s. UU cebn. Zaved. Matematika 5 (6) (1958) 52–90 (Russian).

13. P.C. MQuller, W.O. Schiehlen, Lineare Schwingungen, Akademische Verlagsgesellschaft, Wiesbaden, 1976.
14. C.V. Pao, Logarithmic Derivatives of a Square Matrix, Linear Algebra Appl. 6 (1973) 159–164.
15. C.V. Pao, A further remark on the logarithmic derivatives of a square matrix, Linear Algebra Appl. 7 (1973) 275–278.
16. T. StrQom, Minimization of norms and logarithmic norms by diagonal similarities, Computing 10 (1972) 1–7.
17. T. StrQom, On logarithmic norms, SIAM J. Numer. Anal. 10 (5) (1975) 741–753.
18. W.T. Thomson, Theory of Vibration with Applications, Prentice-Hall, Englewood Cliffs, NJ, 1981.

CITATION

1. L. Kohaupt, Differential calculus for p-norms of complex-valued vector functions with applications, Journal of Computational and Applied Mathematics, Volume 145, Issue 2, 15 August 2002, Pages 425-457, ISSN 0377-0427, http://dx.doi.org/10.1016/S0377-0427(01)00594-5.

Natural Differential Operators and Graph Complexes

Martin Markl
Mathematical Institute of the Academy, Žitná 25,
115 67 Prague 1, the Czech Republic

ABSTRACT

We show how the machine invented by S. Merkulov [S.A. Merkulov, Operads, deformation theory and F-manifolds, in: Frobenius Manifolds, Aspects Math., vol. E36, Vieweg, Wiesbaden, 2004, pp. 213–251; S.A. Merkulov, PROP profile of deformation quantization, Preprint, math. QA/0412257, December 2004; S.A. Merkulov, PROP profile of Poisson geometry, Comm. Math. Phys. 262 (1) (February 2006) 117–135] can be used to study and classify natural operators in differential geometry. We also give an interpretation of graph complexes arising in this context in terms of representation theory. As application, we prove several results on classification of natural operators acting on vector fields and connections.

INTRODUCTION

This work started in an attempt to understand S. Merkulov's idea of "PROP profiles" [18] and [21] and see if and how it may be used to investigate natural structures in geometry. It turned out that classifications of these geometric structures in many interesting cases boiled down to calculations of the cohomology of certain graph complexes.

More precisely, for a wide class of natural operators, the following principle holds.

Principle

For a given type of natural differential operators, there exists a graph cochain complex $(Gr^*, \delta) = Gr^0 \xrightarrow{\delta} Gr^1 \xrightarrow{\delta} Gr^2 \xrightarrow{\delta} \cdots$ such that, in stable ranges,

{Natural operators of a given type} $\cong H_0 (Gr_*, \delta)$.

Stability means that the dimension of the underlying manifold is bigger than some constant explicitly determined by the type of operators. For example, for multilinear natural operators TM×d→TM from the d-fold product of the tangent bundle into itself the stability means that dim (M) $\geq$ d. In smaller dimensions, "exotic" operations described in [5] occur.

In all cases we studied, the corresponding graph complex appeared to be acyclic in positive dimensions, so the cohomology describing natural operators was the only nontrivial piece of the cohomology of (Gr_*, δ). Standard philosophy of strongly homotopy structures [12] suggests that the graph complex (Gr_*, δ) describes stable strongly homotopy operators of a given type.

Graph complexes arising in the Principle are in fact isomorphic to subspaces of fixed elements in suitable Chevalley–Ei lenberg complexes, so, formally speaking, we claim that a certain Chevalley–Eilenberg cohomology is the cohomology of some graph complex. Instances of this phenomenon were systematically used by M. Kontsevich in his seminal paper [9]. The details of operadic graph complexes were then written down by J. Conant [2], J. Conant and K. Vogtmann [3] and [4], M. Mulase and M. Penkava [22], M. Penkava [24], and M. Penkava and A. Schwarz [25]. What makes the Principle exciting is the miraculous fact that the corresponding graph complexes are of the type studied during the "renaissance of operads" and powerful methods developed in this period culminating in [15], [17] and [20]apply.

Another way to view the proposed method is as a formalization of the "abstract tensor calculus" attributed to R. Penrose. When we studied differential geometry in kindergarten, many of us, trying to avoid dozens of indices, drew simple pictures consisting of nodes representing tensors (which resembled little insects) and lines joining legs of these insects symbolizing contraction of indices. We attempt to put this kindergarten approach on a solid footing.

Thus the purpose of this paper is two-fold. The first one is to set up principles of abstract tensor calculus as a useful language for 'stable' geometric objects. This will be done in Sections 1, 2, 3 and 4. The logical continuation should be translating textbooks on differential geometry into this language, because all basic properties of fundamental objects (vector fields, forms, currents, connections and their torsions and curvatures) are of stable nature.

We then show, in Sections 5, 6 and 7, how results on graph complexes may give explicit classifications of natural operators in stable ranges. As an example we derive from a rather deep result of [14] a characterization of operators on vector fields (Theorem 5.1 and its Corollary 5.3). As another application we prove that all natural operators on linear connections and vector fields, with values in vector fields, are freely generated by compositions of covariant derivatives and Lie brackets, and by traces of these compositions—see Theorem 7.2 and Theorem 7.6, and their Corollary 7.3 and Corollary 7.7, in conjunction with Theorem 6.2 and Theorem 6.3.

This article is supplemented by [11] in which we explain the relation between invariant tensors and graphs. We believe that [11], which can be read independently, will help to understand the constructions of Sections 3 and 4.

The theory of invariant operators sketched out in this paper leads to directed, not necessarily connected or simply-connected, graphs. A similar theory can be formulated also for symplectic manifolds, where the corresponding graph complexes would be those appearing in the context of anti-modular operads (modular versions of anticyclic operads, see [16, Definition 5.20]). Something close to a symplectic version of our theory has in fact already been worked out in [27].

NATURAL OPERATORS

Informally, a natural differential operator is a recipe that constructs from a geometric object another one, in a natural fashion, and which is locally a function of coordinates and their derivatives.

Example 1.1

Let M be a n-dimensional smooth manifold. The classical Lie bracket $X,Y \mapsto [X,Y]$ is a natural operation that constructs from two vector fields on M a third one. Given a local coordinate system $(x^1,...,x^n)$ on M, the vector fields X and Y are locally expressions $X = \sum_{1 \le i \le n} X^i \partial / \partial x^i, Y = \sum_{1 \le i \le n} Y^i \partial / \partial x^i$, where X^i, Y^i are smooth functions on M. If we define $X^i_j := \partial X^i / \partial x^j$ and $Y^i_j := \partial Y^i / \partial x^j, 1 \le i, j \le n$ then the Lie bracket is locally given by the formula $[X,Y] = \sum_{1 \le i, j \le n} (X^j Y^i_j - Y^j X^i_j) \partial / \partial x^i$.

In the rest of the paper, we use Einstein's convention assuming summations over repeated indices. In this context, indices i, j,k... will always be natural numbers between 1 and the dimension of the underlying manifold, which will typically be denoted n.

Example 1.2

The covariant derivative $(\Gamma,X,Y) \mapsto \nabla XY$ is a natural operator that constructs from a linear connection Γ and vector fields X and Y , a vector field ∇XY. In local coordinates,

$$\nabla_X Y = \left(\Gamma^i_{jk} X^j Y^k + X^j Y^i_j \right) \frac{\partial}{\partial x^i}$$

$$(1)$$

where Γ^i_{jk} are Christoffel symbols.

Natural operations can be composed into more complicated ones. Examples of 'composed' operations are the torsion $T(X,Y):=\nabla XY-\nabla YX-[X,Y]$ and the curvature $R(X,Y)Z:=\nabla[X,Y]Z-[\nabla X,\nabla Y]Z$ of the linear connection Γ.

Example 1.3

Let X be a vector field and ω a 1-form on M. Denote by $\omega(X) \in C^\infty(M)$ the evaluation of the form ω on X. Then $(X,\omega) \mapsto \exp(\omega(X))$ defines a natural differential operator with values in smooth functions. Clearly, the exponential can be replaced by an arbitrary smooth function $\varphi : \mathbb{R} \to \mathbb{R}$, giving rise to a natural operator $\mathfrak{D}_\varphi(X,\omega) : \varphi(\omega(X))$.

Example 1.4

'Randomly' generated local formulas need not lead to natural operators. As we will see later, neither $0_1(X,y) = X_3^1 Y^4 \partial / \partial x^2$ nor $0_2(X,y) = X^j Y^i_j \partial / \partial x^i$ behaves properly under coordinate changes, so they do not give rise to vector-field valued natural operators.

We may summarize the above examples by saying that a natural differential operator is a recipe given locally as a smooth function in coordinates and their derivatives, such that the local formula is invariant under coordinate changes. After this motivation, we give precise definitions of geometric objects and operators between them. Our exposition follows [7], see also [8].

Denote by Man_n the category of n-dimensional manifolds and open embeddings. Let Fib_n be the category of smooth fiber bundles over n-dimensional manifolds with morphisms differentiable maps covering morphisms of their bases in Man_n.

Definition 1.5

A natural bundle is a functor $B: \text{Man}_n \to \text{Fib}_n$ such that for each $M \in \text{Man}_n$, $\mathfrak{B}(M)$ is a bundle over M. Moreover, $\mathfrak{B}(M')$ is the restriction of $\mathfrak{B}(M)$ for each open submanifold $M' \subset M$, the map $\mathfrak{B}(M') \to \mathfrak{B}(M)$ induced by $M' \hookrightarrow M$ being the inclusion $\mathfrak{B}(M') \hookrightarrow \mathfrak{B}(M)$.

Let us recall a structure theorem for natural bundles due to Krupka, Palais and Terng [10], [23] and [26]. For each s1 we denote by $GL_n^{(s)}$ the group of s-jets of local diffeomorphisms $\mathbb{R}^n \to \mathbb{R}^n$ at 0, so that $GL_n^{(1)}$ is the ordinary general linear group GL_n of linear invertible maps A: $\mathbb{R}^n \to \mathbb{R}^n$. Let $Fr^{(s)}(M)$ be the bundle of s-jets of frames on M whose fiber over

$z \in M$ consist of s -jets of local diffeomorphisms of neighborhoods of $0 \in \mathbb{R}^n$ with neighborhoods of $z \in M$. It is clear that $Fr^{(s)}(M)$ is a principal $GL_n^{(S)}$ -bundle and $Fr^{(1)}(M)$ the ordinary GL_n-bundle of frames $Fr(M)$.

Theorem 1.6 Krupka, Palais, Terng

For each natural bundle B, there exists $l1$ and a manifold $\mathcal{B}$ with a smooth $GL_n^{(l)}$ -action such that there is a functorial isomorphism

$$\mathcal{B}(M) \cong Fr^{(l)}(M) \times_{GL_n^{(l)}} \mathcal{B} := \left(Fr^{(l)}(M) \times \mathcal{B}\right)/GL_n^{(l)} \tag{2}$$

Conversely, each smooth $GL_n^{(l)}$ -manifold $\mathcal{B}$ induces, via (2), a natural bundle $\mathcal{B}$. We will call $\mathcal{B}$ the fiber of the natural bundle $\mathcal{B}$. If the action of $GL_n^{(l)}$ on B does not reduce to an action of the quotient $GL_n^{(l-1)}$ we say that $\mathcal{B}$ has order l.

Example 1.7

Vector fields are sections of the tangent bundle $T(M)$. The fiber of this bundle is $\mathbb{R}^n$, with the standard action of GL_n. The description $T(M) \cong Fr(M)_{GL_n} \times \mathbb{R}^n$ is classical.

Example 1.8

De Rham m -forms are sections of the bundle $\Omega^m(M)$ whose fiber is the space of anti-symmetric m -linear maps $Lin(\Lambda^m(\mathbb{R}^n), \mathbb{R})$, with the obvious induced GL_n-action. The presentation $\Omega^m(M) \cong Fr(M) \times GL_n Lin(\Lambda^m(\mathbb{R}^n), \mathbb{R})$ is also classical. A particular case is $\Omega^0(M) \cong Fr(M)_{GL_n} \times \mathbb{R} \cong M \times R$, the bundle whose sections are smooth functions. We will denote this natural bundle by $\mathfrak{R}$, believing there will be no confusion with the symbol for the reals.

Example 1.9

Linear connections are sections of the bundle of connections $Con(M)$ [8, Section 17.7] which we recall below. Let us first describe the group $GL_n^{(2)}$. Its elements are expressions of the form $A=A_1+A_2$, where $A_1 : \mathbb{R}$

$^n \to \mathbb{R}^n$ is a linear invertible map and A_2 is a linear map from the symmetric product $\mathbb{R}^n \odot \mathbb{R}^n$ to $\mathbb{R}^n$. The multiplication in $GL_n^{(2)}$ is given by

$$(A_1 + A_2)(B_1 + B_2) := A_1(B_1) + A_1(B_2) + A_2(B_1, B_1).$$

The unit of $GL_n^{(2)}$ is $id_{\mathbb{R}^n} + 0$ and the inverse is given by the formula

$$(A_1 + A_2)^{-1} = A_1^{-1} - A_1^{-1}(A_2(A_1^{-1}, A_1^{-1}))$$

Let C be the space of linear maps $Lin(\mathbb{R}^n \otimes \mathbb{R}^n, \mathbb{R}^n)$, with the left action of $GL_n^{(2)}$ given as

$$(Af)(u \otimes v) := A_1 f(A_1^{-1}(u), A_1^{-1}(v)) - A_2(A_1^{-1}(u), A_1^{-1}(v)) \tag{3}$$

for $f \in Lin(\mathbb{R}^n \otimes \mathbb{R}^n, \mathbb{R}^n)$, $A = A_1 + A_2 \in GL_n^{(2)}$ and $u, v \in \mathbb{R}^n$. The bundle of connections is then the order 2 natural bundle represented as $Con(M) := Fr^{(2)}(M) \times_{GL^{(2)}} C$. Observe that, while the action of $GL^{(2)}$ on the vector space C is not linear, the restricted action of $GL_n \subset G_n^{(2)}$ on C is the standard action of the general linear group on the space of bilinear maps.

For $k \geq 0$ we denote by $\mathfrak{B}^{(k)}$ the bundle of k-jets of local sections of the natural bundle $\mathfrak{B}$ so that $\mathfrak{B}^{(0)} = \mathfrak{B}$. If B is represented as in (2), then $\mathfrak{B}^{(k)}(M) \cong Fr^{(k+l)}(M) \times_{GL^{(k+l)}} B^{(k)}$, where $B^{(k)}$ is the space of k-jets of local diffeomorphisms $\mathbb{R}^n \to B$ defined in a neighborhood of $0 \in \mathbb{R}^n$.

Definition 1.10

Let $\mathfrak{F}$ and $\mathfrak{G}$ be natural bundles. A (finite order) natural differential operator $\mathfrak{D}: \mathfrak{F} \to \mathfrak{G}$ is a natural transformation (denoted by the same symbol) $\mathfrak{D}: \mathfrak{F}^{(k)} \to \mathfrak{G}$, for some k1. We denote the space of all natural differential operators $\mathfrak{F} \to \mathfrak{G}$ by $\mathfrak{Nat}(\mathfrak{F}, \mathfrak{G})$.

If $\mathfrak{F}$ and $\mathfrak{G}$ are natural bundles of order l, with fibers F and G, respectively, then each natural operator in Definition 1.10 is induced by an GL_n^{k+l}-equivariant map $0: F^{(k)} \to G$, for some k0. Conversely, such an equivariant map induces an operator $\mathfrak{D}: \mathfrak{F} \to \mathfrak{G}$. This means that the study of natural operators is reduced to the study of equivariant maps. The procedure described above is therefore called the IT reduction (from invariant-theoretic).

From this moment on, we impose the following assumptions on natural bundles $\mathfrak{F}$, $\mathfrak{G}$ and operators $\mathfrak{O}: \mathfrak{F} \to \mathfrak{G}$ between them.

A1: The fibers $\mathcal{F}$ and $\mathcal{G}$ of the bundles $\mathfrak{F}$ and $\mathfrak{G}$ are vector spaces and the restricted actions of $GL_n \subset GL_n^{(l)}$ on $\mathcal{F}$ and $\mathcal{G}$ are rational linear representations,

A2: The action of $GL_n^{(l)}$ on the fiber $\mathcal{G}$ of $\mathfrak{G}$ is linear, and

A3: We consider only polynomial differential operators for which the induced map of the fibers $O: \mathcal{F}^{(k)} \to \mathcal{G}$ is a polynomial map.

Notice that we do not require the action of the full group $GL_n^{(l)}$ on the fiber of F to be linear. Assumption A_2 is needed for the cohomology in Theorem 2.2 in Section 2 to be well-defined, assumptions A_1 and A_3 are necessary to relate this cohomology to a graph complex.

Polynomiality A_3 rules out operators as $\mathfrak{O}_\varphi$ from Example 1.3. There is probably no systematic way how to study operators of this type— imagine that φ is an arbitrary, not even real analytic, smooth function. Clearly most if not all "natural" natural operators considered in differential geometry are polynomial, so assumption A3 seems to be justified. As argued in [8, Section 24] and as we will also see later in Remark 5.2 and Remark 7.1, in some situations the operators possess a certain homogeneity which automatically implies polynomiality.

Example 1.11

Given natural bundles $\mathfrak{B}'$ and $\mathfrak{B}''$ with fibers B' resp. B'', there is an obviously defined natural bundle $\mathfrak{B}' \times \mathfrak{B}''$ with fiber $B' \times B''$. With this notation, the Lie bracket is a natural operator $[-,-]: T \times T \to T$ and the covariant derivative an operator $\nabla: Con \times T \times T \to T$, where T is the tangent space functor and Con the bundle of connections recalled in Example

1.9. The corresponding equivariant maps of fibers can be easily read off from local formulas given in Example 1.1 and Example 1.2.

Example 1.12

The operator $\mathfrak{D}_\varphi : T \times \Omega^1 \to C^\infty$ from Example 1.3 is induced by the GL_n-equivariant map $O_\varphi : \mathbb{R}^n \times \mathbb{R}^{n*} \to \mathbb{R}$ given by $O_\varphi(v, \alpha) := \varphi(\alpha(v))$. Clearly, O_φ satisfies A_3 if and only if $\varphi : \mathbb{R} \to \mathbb{R}$ is a polynomial.

NATURAL OPERATORS AND COHOMOLOGY

We start this section by a brief recollection of two classical constructions. For a Lie algebra $\mathfrak{h}$ and a $\mathfrak{h}$-module W, the Chevalley–Eilenberg cohomology $H^*(\mathfrak{h}, W)$ of h with coefficients in W is the cohomology of the cochain complex $(C^*(\mathfrak{h}, W), \delta_{CE})$ defined by

$$C^m(\mathfrak{h}, W) := \lin(\wedge^m \mathfrak{h}, W), m \geq 0,$$ with δ_{CE} the sum $\delta_{CE} = \delta_1 + \delta_2$, where

$$(\delta_1 f)(h_1, \ldots, h_{m+1}) := \sum_{1 \leq i \leq m+1} (-1)^{i+1} \cdot h_i f(h_1, \ldots, \hat{h}_i, \ldots, h_{m+1})$$
(4)

$$(\delta_2 f)(h_1, \ldots, h_{m+1}) := \sum_{1 \leq i < j \leq m+1} (-1)^{i+j} \cdot f([h_i, h_j], h_1, \ldots, \hat{h}_i, \ldots, \hat{h}_j, \ldots, h_{m+1})$$
(5)

for $f \in C^m(\mathfrak{h}, W)$, $h_1, \ldots, h_{m+1} \in \mathfrak{h}$ and $\wedge$ denoting the omission. If $m = 0$, the summation in the right-hand side of (5) runs over the empty set, so we put $(\delta_2 f)(h) := 0$ for $f \in C^0(\mathfrak{h}, W)$.

The second notion we need to recall is the semidirect product of groups. Assume that G and H are Lie groups, with G acting on H by homomorphisms. One then defines the semidirect product $G \ltimes H$ as the Cartesian product $G \times H$ with the multiplication

$$(g_1, h_1)(g_2, h_2) := (g_1 g_2, g_2^{-1}(h_1) h_2), \quad g_1, g_2 \in G, \ h_1, h_2 \in H.$$

Both G and H are subgroups of G⋊H and their union G∪H generates G⋊H. Let us close this introductory part by formulating a proposition that ties the above two constructions together.

If W is a left G⋊H-module, the inclusion H⊂G⋊H induces a left H-action on W which in turn induces an infinitesimal action of $\mathfrak{h}$ on W. One may therefore consider the cochain complex $(C^*(\mathfrak{h},W),\delta_{CE})$. Since G acts by homomorphisms, the unit of H is G-fixed, so there is an induced action of G on the Lie algebra $\mathfrak{h}$ of H. The group G acts also on W, via the inclusion G⊂G⋊H. These two actions give rise, in the usual way, to an action of G on $C^*(\mathfrak{h},W)$. Let us denote $C_G^*(\mathfrak{h},W)$ the subspace of G -fixed elements of $C^*(\mathfrak{h},W)$. We have the following:

Proposition 2.1

The subspace of fixed elements $C_G^*(\mathfrak{h},W)\subset C^*(\mathfrak{h},W)$ is δ_{CE}-closed, so the cohomology $H_G^*(\mathfrak{h},W):=H^*\left(C_G^*(\mathfrak{h},W),\delta_{CE}\right)$ is defined. For H connected, there is an isomorphism

$$H_G^0(\mathfrak{h},W)\cong W^{G\times H} \tag{6}$$

where $W^{G\rtimes H}$ denotes, as usual, the space of G⋊H-fixed elements in W.

Proof: We leave a direct verification of the δ_{CE}-closeness of $C_G^*(\mathfrak{h},W)$ as a simple exercise to the reader. It is equally easy to see that $H_G^0(\mathfrak{h},W)$ consists of elements of W which are simultaneously G -fixed and $\mathfrak{h}$-invariant. If H is connected, the exponential map is an epimorphism, thus $\mathfrak{h}$-invariant elements in W are precisely those which are H -fixed. This, along with the fact that G∪H generates G⋊H, gives (6).

In Section 1 we recalled that natural differential operators $\mathcal{D}\in\mathfrak{Nat}(\mathcal{F},\mathcal{G})$ between natural bundles of order l with fibers $\mathcal{F}$ resp. $\mathcal{G}$, correspond to $G_n^{(k+l)}$-equivariant maps $0:\mathcal{F}^{(k)}\to\mathcal{G}$ with some k0. This can be expressed by the isomorphism:

$$\mathfrak{Nat}(\mathfrak{F}, \mathfrak{G}) \cong \bigcup_{k \geqslant 0} Map_{GL_n^{(k+l)}}(\mathcal{F}^{(k)}, \mathcal{G})$$

$$(7)$$

where $Map_{GL_n^{(k+l)}}(\mathcal{F}^{(k)}, \mathcal{G})$ is the space of polynomial $GL_n^{(k+l)}$-equivariant maps $F^{(k)} \to G$—see assumption A3 on page 260. The space $Map(F^{(k)}, G)$ of all polynomial maps has the standard $GL_n^{(k+l)}$-action induced from the actions on $F^{(k)}$ and G.

The space of equivariant maps is the fixed subspace $Map_{GL_n^{(k+l)}}(\mathcal{F}^{(k)}, \mathcal{G}) = Map(\mathcal{F}^{(k)}, \mathcal{G}) \, GL_n^{(k+l)}$

. Let us see how Proposition 2.1 describes these spaces. The crucial observation is that $GL_n^{(S)}$ is, for each s1, a semidirect product [8, Section 13]. If $(\mathbb{R}^n)^{\odot r}$ denotes the rth symmetric power of $\mathbb{R}^n$, r≥1, then elements of $GL_n^{(S)}$ are expressions $A = A_1 + A_2 + A_3 + \cdots + A_s$, $A_i \in Lin((\mathbb{R}^n)\odot i, \mathbb{R}^n)$, 1is, such thatA1: $\mathbb{R}^n \to \mathbb{R}^n$ is invertible. It is a simple exercise to write formulas for the product and inverse; for s=2 it was done in Example 1.9.

The space $Lin((\mathbb{R}^n)\odot i, \mathbb{R}^n)$ is canonically isomorphic to the space $Sym((\mathbb{R}^n)^{\otimes i}, \mathbb{R}^n)$ of symmetric multilinear maps and we will identify these two spaces in the sequel. Denote by $NGL_n^{(S)} = \{A = A_1 + A_2 + Al \cdots + A_s \varepsilon NGL_n^{(S)} A_1 = id\}$ the prounipotent radical of $GL_n^{(S)}$. Then $GL_n^{(S)}$ is the semidirect product $GL_n^{(S)} = GL_n \times NGL_n^{(S)}$, with GL_n acting on $NGL_n^{(S)}$ by adjunction. Denote finally $ngr_n^{(S)}$ the Lie algebra of $NGL_n^{(S)}$,

$$\mathfrak{ngl}_n^{(S)} = \{a = a_2 + a_3 + \cdots + a_s; \; a_i \in Sym((\mathbb{R}^n)^{\otimes i}, \mathbb{R}^n), \; 2 \leqslant i \leqslant s\}$$

$$(8)$$

Assume that the action of $Gl_n^{(l)}$ on the fiber $\mathcal{G}$ of $\mathfrak{G}$ is linear. Then Map($\mathcal{F}^{(k)}, \mathcal{G}$) is a linear representation of Gl_n^{k+l} and Proposition 2.1 applied to G=GLn, H = NGL_n^{k+l} and W=Map(F($\mathcal{F}^{(k)}, \mathcal{G}$)) gives

$$Map_{GL_n^{(k+l)}}(\mathcal{F}^{(k)}, \mathcal{G}) \cong H^0_{GL_n}(\mathfrak{ngl}_n^{(k+l)}, Map(\mathcal{F}^{(k)}, \mathcal{G}))$$

$$(9)$$

For each k0, the inclusion Map($\mathcal{F}^{(k)}, \mathcal{G}$)$\hookrightarrow$Map($\mathcal{F}^{(k+1)}, \mathcal{G}$) together with

the projection $\mathfrak{ngr}_n^{(k+l+1)} \to \mathfrak{ngr}_n^{(k+l)}$ induces a GL_n-invariant inclusion

$$C^*(\mathfrak{ngl}_n^{(k+l)}, Map(\mathcal{F}^{(k)}, \mathcal{G})) \hookrightarrow C^*(\mathfrak{ngl}_n^{(k+l+1)}, Map(\mathcal{F}^{(k+1)}, \mathcal{G}))$$

which commutes with the differentials. Let us denote

$$C^*(\mathfrak{ngl}_n^{(\infty)}, Map(\mathcal{F}^{(\infty)}, \mathcal{G})) := \bigcup_{k \geqslant 0}\bigcup_{l \geqslant 1} C^*(\mathfrak{ngl}_n^{(k+l)}, Map(\mathcal{F}^{(k)}, \mathcal{G}))$$

$$(10)$$

and $C^*_{GL_n}(\mathfrak{ngr}_n^{(\infty)}, Map(\mathcal{F}^{(\infty)}, \mathcal{G}))$ the GL_n-stable subspace of C^* (

$\mathfrak{ngr}_n^{(\infty)}, Map(\mathcal{F}^{(\infty)}, \mathcal{G}))$. Let

$$H^*_{GL_n^{(\infty)}}(\mathfrak{ngl}_n^{(\infty)}, Map(\mathcal{F}^{(\infty)}, \mathcal{G})) := H^*(C^*_{GL_n}(\mathfrak{ngl}_n^{(\infty)}, Map(\mathcal{F}^{(\infty)}, \mathcal{G})), \delta_{CE})$$

.

Then (7) together with (9) and the fact that cohomology commutes
with direct limits implies:

Theorem 2.2: Let $\mathfrak{F}$ and $\mathfrak{G}$ be natural bundles with fibers $\mathcal{F}$ resp. $\mathcal{G}$
of orders l. Suppose that the action of $GL_n^{(l)}$ on $\mathcal{G}$ is linear. Then, under
the above notation

$$\mathfrak{Nat}(\mathfrak{F}, \mathfrak{G}) \cong H^0_{GL_n}(\mathfrak{ngl}_n^{(\infty)}, Map(\mathcal{F}^{(\infty)}, \mathcal{G}))$$

$$(11)$$

In the following sections we show that, in many interesting cases, the
cohomology in the right-hand side of (11) is the cohomology of a certain graph complex.

NATURAL OPERATORS AND GRAPHS

We are going to describe natural differential operators by certain spaces spanned by graphs. Roughly speaking, graphs, viewed as contraction schemes for indices, will encode elementary GL_n-invariant tensors in (10). Our approach is based on a translation of the Invariant Tensor Theorem into the graph language explained in [11]

Suppose that $\mathfrak{B}$ is a natural bundle satisfying A1 on page 260, so that the induced action of $GL_n \subset GL_n^{(l)}$ on the fiber B is rational linear. According to standard facts of the representation theory of GL_n recalled, for instance, in [7, § 1.4], an equivalent assumptions is that, as a GL_n-module, B is the direct sum of GL_n-modules

$$\mathfrak{B} = \bigoplus_{1 \leqslant i \leqslant b} \mathfrak{B}_i$$

(12)

where $\mathfrak{B}_i$ is, for each 1ib, either the space $\text{Lin}(\mathbb{R}^{n\otimes q_i}, \mathbb{R}^{n\otimes p_i})$ for some pi,qi0, with the standard GL_n-action, or a subspace of this space consisting of maps whose inputs and/or outputs have a specific symmetry, which can for example be expressed by a Young diagram.

In other words, $\mathcal{B}_i$ are spaces of multilinear maps whose coordinates are tensors $T_{b_1,\cdots,b_{q_i}}^{a_1,\cdots,a_{p_i}}$ with q_i input indices and p_i output indices, which may or may not enjoy some kind of symmetry. We will graphically represent these tensors as corollas with q_i-inputs and p_i outputs:

p_i outputs

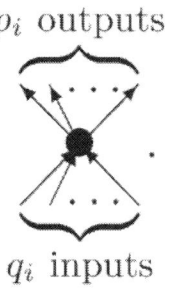

q_i inputs

(13)

Instead of • we may sometime use different symbols for the node, such as ▽, ■, ○, &c.

Example 3.1

The fiber of the tangent bundle T is $\mathbb{R}^n = \text{Lin}(\mathbb{R}^{n\otimes 0}, \mathbb{R}^{n\otimes 1})$, so one has in (12) b=1, $p_1=1$, $q_0=0$. Elements of the fiber are tensors X^a symbolized by •. The fiber $\mathcal{C}$ of the connection bundle Con (see Example 1.9) is Lin($\mathbb{R}^{n\otimes 2}$, $\mathbb{R}^{n\otimes 1}$), therefore b=1, $p_1=1$ and $q_1=2$. Elements of $\mathcal{C}$ are GL_n-tensors (Christoffel symbols) Γ^a_{bc} represented by

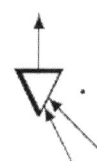

An example with a(n anti-)symmetry is the bundle Ω^m of de Rham m-forms, m0. Its fiber is the space $\text{Lin}(\Lambda^m \mathbb{R}^n, \mathbb{R}^{n\otimes 0)} = \text{Lin}(\Lambda^m \mathbb{R}^n, \mathbb{R})$ of anti-symmetric tensors $\omega)b_1,\ldots,b_m($.

Sometimes we will need decorations of nodes. For example, the product bundle T×T has fiber $\mathbb{R}^n \times \mathbb{R}^n$ generated by tensors X^a, Y^a which will be denoted

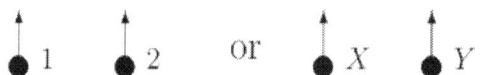

Let $\mathfrak{B}$ be a natural bundle with fiber $\mathcal{B}$ decomposed as in (12). It is easy to see that the fiber $\mathcal{B}^{(k)}$ of the k-jet bundle $\mathfrak{B}^{(k)}$ decomposes, as a GL_n-module, into $\mathcal{B}^{(k)} = \oplus_{1\leq i\leq b} \mathcal{B}_i^{(k)}$, where

$$\mathcal{B}_i^{(k)} = \bigoplus_{0\leq v\leq k} Sym(\mathbb{R}^{n\otimes v}, \mathbb{R}) \otimes \mathcal{B}_i$$

(14)

This means that if elements of $\mathcal{B}_i$ are tensors $T^{a_1,\dots,a_{p_i}}_{b_1,\dots,b_{q_i}}$, elements of $\mathcal{B}^{(k)}_i$ are tensors $_{(s_1,\dots,s_v)}T^{a_1,\dots,a_{p_i}}_{b_1,\dots,b_{q_i}}$, vk, with braces indicating the symmetry in $(s_1,\dots,s_v)$. In terms of pictures this amounts to adding new symmetric inputs to corollas (13), so elements of $\mathcal{B}^{(k)}_i$ will be symbolized by

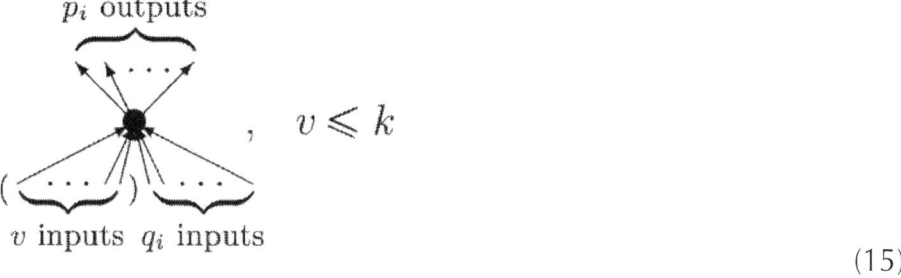

$$v \leqslant k \tag{15}$$

Example 3.2
The fiber of the k th tangent bundle $T^{(k)}$ is the space of tensors

$$X^a_{(s_1,\dots,s_v)} := \frac{\partial^u X^a}{\partial x^{s_1}\dots\partial x^{s_v}}, \quad v \leqslant k \tag{16}$$

which we draw as

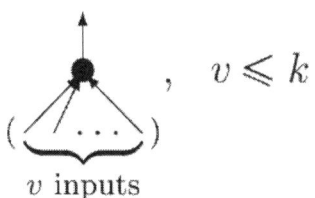

$$v \leqslant k \tag{17}$$

The fiber of the bundle $\mathrm{Con}^{(k)}$ is the space of tensors $_{(s_1,\dots,s_v)}\Gamma^a_{bc} := \dfrac{\partial^u \Gamma^a_{bc}}{\partial x^{s_1}\dots\partial x^{s_v}}$, vk, depicted as

$$v \leqslant k$$

v inputs

(18)

As follows from (8), $ngr_n^{k+1} = \oplus_{2 \leq u \leq k+l} Sym(\mathbb{R}^{n \otimes u}, \mathbb{R}^n)$. Therefore ngl_n^{k+1} is the space of symmetric tensors $\varphi_{(s_1,\ldots,s_u)}^b$, $2uk+l$, or in pictures,

$$2 \leqslant u \leqslant k + l$$

u inputs

(19)

In what follows, white corollas (19) will always denote   elements of $ngl_n^{(k+1)}$ for some k+l2.

In the rest of this section we construct a graded space $\mathcal{Gr}^*_{\mathfrak{F},\mathfrak{G}}$ spanned by graphs representing GL_n-invariant cochains in $C^*(ngl_n^{(\infty)}, Map(\mathcal{F}^{(\infty)}, \mathcal{G}))$. The differentials will be studied in the next section.

Suppose that the natural bundles $\mathfrak{F}$ and $\mathfrak{G}$ satisfy assumption A1 on page 260, and see what can be said about the space $C_{GL_n}^m(ngr_n^{k+1}, Map(\mathcal{F}^{(k)}, \mathcal{G})$ of GL_n-equivariant polynomial maps from $Map(ngr_n^{(k+1)} \times \mathcal{F}^{(k)}, \mathcal{G})$ that are m-homogeneous and antisymmetric in $ngl_n^{(k+l)}$. By the polynomiality assumption A3,

$$C_{GL_n}^m\left(ngl_n^{(k+l)}, Map(\mathcal{F}^{(k)}, \mathcal{G})\right) \cong \bigoplus_{t \geq 0} Lin_{GL_n}\left(\bigwedge^m ngl_n^{(k+l)} \otimes \mathcal{F}^{(k) \otimes t}, \mathcal{G}\right)$$

(20)

where $LinGL_n(-,-)$ denotes the space of GL_n-equivariant linear maps.

Let us decompose the fibers F and G of natural bundles $\mathfrak{F}$ and $\mathfrak{G}$ into the direct sum (12), $\mathcal{F} = \oplus 1 \leqslant i \leqslant f^{\mathcal{F}}$ and $\mathcal{G} = \oplus 1 \leqslant i \leqslant g^{\mathcal{G}}$. By (14), the components of the fiber $\mathfrak{F}^{(k)}$ of the k-jet bundle $\mathfrak{F}^{(k)}$, $k \geqslant 0$, are the direct sums $\mathcal{F}_i^{(k)} = \oplus_{0 \leqslant v \leqslant k} \mathcal{F}_i^{[v]}$, 1if, with $\mathcal{F}_i^{[v]} := \mathrm{Sym}(\mathbb{R}^{n \otimes v}, \mathbb{R}^n) \oplus \mathcal{F}$. Using the above decompositions and description (8) of $\mathrm{ngr}_n^{(k+1)}$, one can rewrite the right-hand side of (20) into

$$\bigoplus_{t \geqslant 0} \bigoplus_{S(k,l,t)} \mathrm{Lin}_{GL_n}\left(\bigwedge_{1 \leqslant i \leqslant m} \mathrm{Sym}(\mathbb{R}^{n \otimes u_i}, \mathbb{R}^n) \otimes \bigotimes_{1 \leqslant s \leqslant t} \mathcal{F}_{i_s}^{[v_s]}, \mathcal{G}_i \right)$$

(21)

where S(k,l,t) is the set of integers $u_1, \ldots, u_m,\ i_1, \ldots, i_t,\ v_1, \ldots, v_t$ and i such that

$$2 \leqslant u_1, \ldots, u_m \leqslant k+l, \quad 1 \leqslant i_1, \ldots, i_t \leqslant f, \quad 0 \leqslant v_1, \ldots, v_t \leqslant k \quad \text{and} \quad 1 \leqslant i \leqslant g$$

Let us fix a multiindex $\omega = (u_1, \ldots, u_m, i_1, \ldots, i_t, v_1, \ldots, v_t, i) \in S(k,l,t)$. By our assumptions, the space $\mathcal{F}_{i_s}^{[v_s]}$ is, for each 1st, isomorphic to the space

$$\mathrm{Lin}_{\mathfrak{J}_s}^{\mathfrak{D}_s}(\mathbb{R}^{n \otimes (v_s + q_{i_s})}, \mathbb{R}^{n \otimes p_{i_s}}) := \{f \in \mathrm{Lin}(\mathbb{R}^{n \otimes (v_s + q_{i_s})}, \mathbb{R}^{n \otimes p_{i_s}}); \ fs = 0 = tf \text{ for } s \in \mathfrak{J}_s, t \in \mathfrak{D}_s\}$$

of linear maps having a symmetry specified by subsets $\mathfrak{J}_s \subset k[\Sigma_{v_s + q_{i_s}}], \mathfrak{D}_s[\Sigma p_{i_s}]$, see also [11, Remark 4.4]. Similarly, $\mathcal{G}_i \cong \mathrm{Lin}_{\mathfrak{J}}^{\mathfrak{D}}(\mathbb{R}^{n \oplus c}, \mathbb{R}^{n \oplus d})\mathfrak{J}$.For some c,d0 and subsets $\mathfrak{J} \subset k[\Sigma_c], \mathfrak{D} \subset k[\Sigma_d]$. The expression

$$\mathrm{Lin}_{GL_n}\left(\bigwedge_{1 \leqslant i \leqslant m} \mathrm{Sym}(\mathbb{R}^{n \otimes u_i}, \mathbb{R}^n) \otimes \bigotimes_{1 \leqslant s \leqslant t} \mathcal{F}_{i_s}^{[v_s]}, \mathcal{G}_i \right)$$

(22)

in (21) is therefore isomorphic to

$$Lin_{GL_n}\left(\bigwedge_{1\leqslant i\leqslant m} Sym(\mathbb{R}^{n\otimes u_i},\mathbb{R}^n)\otimes \bigotimes_{1\leqslant s\leqslant t} Lin_{\mathfrak{I}_s}^{\mathfrak{D}_s}(\mathbb{R}^{n\otimes(v_s+q_{is})},\mathbb{R}^{n\otimes p_{is}}), Lin_{\mathfrak{I}}^{\mathfrak{D}}\right.$$

$$\left.(\mathbb{R}^{n\otimes c},\mathbb{R}^{n\otimes d})\right)$$

$$(23)$$

Let us remark that in all applications discussed in this paper, we will always have $pi_s=1$ for 1st, $c=0$ and $d=1$.

Observe that (23) is the space in (24) of [11], with an appropriate choice of the parameters, which in this case is $r:=t-m$, and

$$h_i := u_i \qquad\qquad\qquad \text{for } 1\leqslant i\leqslant m, \text{ and}$$

$$h_i := v_s + q_{i_s}, \quad \mathfrak{I}_i := \mathfrak{I}_s, \quad \mathfrak{D}_i := \mathfrak{D}_s \quad \text{for } i=s+m, \ 1\leqslant s\leqslant t$$

therefore the methods developed in [11] apply. We believe that the reader can tolerate a certain incompatibility between the notation used in this paper and the notation of [11]—the alphabet does not have enough letters to avoid notational conflicts.

By Proposition 4.8 and Remark 4.10 of [11], the space (23) is related to the space $\mathcal{G}\Gamma_{\omega}^m$ spanned by graphs with vertices of three types:

1st type: t 'black' vertices (15) with $pi:=pi_s$, $qi:=qi_s$ and $v:=v_s$, representing tensors in $\mathcal{F}_{j_s}^{[v_s]}$, 1st,

2nd type: one vertex (13) with $p_i: =c$ and $q_i:=d$ called the anchor, representing tensors in the dual $\mathcal{G}_i^*$ of $\mathcal{G}_i$.

3rd type: m 'white' vertices (19) with $u:=u_i$ representing generators of the Lie algebra $\mathfrak{ngl}_n^{(k+1)}$, 1im.

Our graphs are directed and oriented, where an orientation is, by definition, an equivalence class of linear orders of the set of white vertices, modulo the relation identifying orders that differ by an even number of transpositions. If the orientations of two graphs G' and G'' differ by an odd number of transpositions, we put $G' = -G''$ in $\mathcal{G}\Gamma^m_\omega$

This notion of orientation is not the traditional one but resembles orientations in various graph complexes [16, § II.5.5].

The graphs spanning $\mathcal{G}\Gamma^m_\omega$ are not required to be connected, and multiple edges and loops are allowed. The vertices above are Merkulov's genes [21]. The unique vertex of the 2nd type marks the place where we evaluate the composition along the graph at an element of $\mathcal{G}^*$, which explains the dualization in the definition of this vertex.

Proposition 4.8 of [11] (or its obvious extension mentioned in [11, Remark 4.10]), combined with the isomorphism between (22) and (23), gives an epimorphism

$$R^m_{n,\omega} : \mathcal{G}r^m_\omega \twoheadrightarrow Lin_{GL_n}\left(\bigwedge_{1 \leqslant i \leqslant m} Sym(\mathbb{R}^{n \otimes u_i}, \mathbb{R}^n) \otimes \bigotimes_{1 \leqslant s \leqslant t} \mathcal{F}^{[v_s]}_{i_s}, \mathcal{G}_i \right)$$

(24)

which is, by [11, Proposition 4.9], a monomorphism if $n+m$ the number of edges of graphs in $\mathcal{G}\Gamma^m_\omega$. The central result of this section, Theorem 3.3 below, uses the limit

$$\mathcal{G}r^m_{\mathcal{F},\mathcal{B}} := \bigcup_{k \geqslant 0} \bigcup_{l \geqslant 1} \bigoplus_{t \geqslant 0} \bigoplus_{\omega \in S(k,l,t)} \mathcal{G}r^m_\omega$$

(25)

The space $\mathcal{G}r^m_{\mathcal{F},\mathcal{B}}$ is spanned by graphs with an arbitrary number of the 1st type vertices with an arbitrary v0 in (15), one 2nd type vertex representing tensors in $\mathcal{G}^*_i$ for 1ig, and m 3rd type vertices with an arbitrary u2 in (19).

Theorem 3.3: The epimorphosis $R^m_{n,\omega}$ in (24) assemble, for each m0, into a surjection

$$R^m_n : \mathcal{G}r^m_{\mathcal{F},\mathcal{B}} \twoheadrightarrow C^m_{GL_n}\left(\mathfrak{ngl}^{(\infty)}_n, Map(\mathcal{F}^{(\infty)}, \mathcal{G}) \right)$$

(26)

The restriction

$$R_n^m(e) : \mathcal{G}r_{\mathcal{F},\mathcal{G}}^m(e) \to C_{GL_n}^m\left(\mathfrak{ngl}_n^{(\infty)}, Map(\mathcal{F}^{(\infty)}, \mathcal{G})\right)$$

of the map R_n^m to the subspace $\mathcal{G}r_{\mathcal{F},\mathcal{G}}^m(e) \subset \mathcal{G}r_{\mathcal{F},\mathcal{G}}^m$ spanned by graphs with e edges, is a monomorphism whenever $n = \dim(M)e - m$.

Proof: The maps $R_{n,\omega}^m$ of (24) assemble, for each k0 and l1, into an epimorphism

$$R_{n,k,l}^m := \bigoplus_{t \geqslant 0} \bigoplus_{\omega \in S(k,l,t)} R_{n,\omega}^m : \bigoplus_{t \geqslant 0} \bigoplus_{\omega \in S(k,l,t)} \mathcal{G}r_\omega^m \twoheadrightarrow \bigoplus_{t \geqslant 0} Lin_{GL_n}\left(\bigwedge^m \mathfrak{ngl}_n^{(k+l)} \otimes \mathcal{F}^{(k) \otimes t}, \mathcal{G}\right).$$

Recalling (10) and (20), and the definition (25) of the graph complex $\mathcal{G}r_{\mathcal{F},\mathcal{G}}^m$, we conclude that $R_n^m := \bigoplus_{k \geqslant 0} \bigoplus_{i \geqslant 1} R_{n,k,l}^m$ is the desired surjection (26). The second part of the theorem follows from [11, Proposition 4.9] applied to the constituents $R_{n,\omega}^m$ of R_n^m.

Example 3.4
Let us discuss the case $\mathcal{F} = T \times T$ and $\mathcal{G} = T$, where T is the tangent bundle functor. Graphs spanning the vector space $\mathcal{G}r_{T \times T \times T}^m$ have finite number of the 1st type vertices (17)

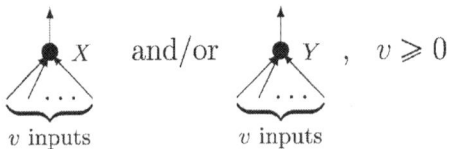

X and/or Y , $v \geqslant 0$

v inputs $\qquad$ v inputs

marking the places where to insert tensors $X_{(S_1,...,Sv)}^a$ and $Y_{(S_1,...,Sv)}^a$ of the fiber of $(T \times T)(\infty)$. The unique vertex ⊺ of the 2nd type is the place to insert a tensor of the fiber $\mathbb{R}^{n*}$ of T^*. There of course will also be m vertices (19) of the 3rd type for generators of $\mathfrak{ngl}_n^\infty$.

Observe that we omitted braces indicating the symmetry because inputs of all vertices are symmetric and no confusion may occur. Let us inspect how $\mathcal{G}r^0_{T\times T,T}$ describes GLn-equivariant maps in

$$\mathrm{Map}_{GL_n}\left(\left(\mathbb{R}^n\times\mathbb{R}^n\right)^{(\infty)},\mathbb{R}^n\right)=C^0_{GL_n}\left(\mathrm{ngr}_n^{(\infty)},\mathrm{Map}\left(\left(\mathbb{R}^n\times\mathbb{R}^n\right)^{(\infty)},\mathbb{R}^n\right)\right).$$ The graph

describes the equivariant map that sends an element $\left(X^a,X^a_b,Y^a,Y^a_b\right)\in$ $\left(\mathbb{R}^n\times\mathbb{R}^n\right)^{(1)}$ into the element $(X^iY^a_j)\in\mathbb{R}^n$. It is precisely the map O_2 considered in Example 1.4. The linear combination

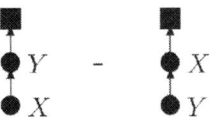

represents the local formula $\left(X^a,X^a_b,Y^a,Y^a_b\right)\mapsto\left(X^iY^a_j-Y^iY^a_j\right)$ for the L ie bracket [X,Y] of two vector fields. We allow also graphs as

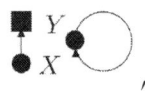

which represents the map $\left(X^a,X^a_b,Y^a,Y^a_b\right)\mapsto\left(X^iY^i_i\right)$ involving the trace Y^i_i of Y . An example of a degree 1 cochain in $C^1_{GL_n}\left(\mathrm{ngr}_n^{(2)},\mathrm{Map}\left(\mathbb{R}^n\times\mathbb{R}^n,\mathbb{R}^n\right)\right)$ is provided by

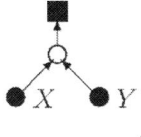

which defines the GLn-equivariant 1-cochain $\left(\varphi^a_{bc},X^a,Y^a\right)\mapsto\left(\varphi^a_{ij}X^iY^j\right)$.

As explained in [11, Remark 5.2], for degrees 2 our interpretation of graphs involves the antisymmetrization in white vertices. For instance, the graph

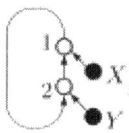

represents the 2-cochain $\left(\varphi^a_{bc},\psi^a_{bc},X^a,Y^a\right)\mapsto\left(\varphi^i_{jk}\psi^j_{il}\right)X^kY^l$. The reason why the expected traditional $\dfrac{1}{2!}$-factor is missing is explained in Remark 4.5.

Example 3.5

In this example we express local formulas for the covariant derivative, torsion and curvature in terms of graphs. The covariant derivative is the operator $\nabla\colon\mathrm{Con}\times T^{\times 2}\to T$ locally given by the graph

$$\nabla_X Y:$$

(27)

which is a graphical form of formula (1). The torsion $T\colon\mathrm{Con}\times T^{\times 2}\to T$ is given by

$$T(X,Y):$$

and the curvature $R\colon\mathrm{Con}\times T^{\times 3}\to T$ as

$$R(X,Y)Z :$$

Example 3.6

This example shows that the map R_n^m from Theorem 3.3 need not be a monomorphism below the 'stable range.' Consider again the two graphs from Example 3.4:

The number of edges of both graphs is 2. As we already saw, G_1 represents the local formula $\sum_{1\le i,j\le n} X^j / \partial Y^i / \partial X^j \partial / \partial x^i$ and G2 the formula $\sum_{1\le i,j\le n} \partial Yj / \partial xj X \partial / \partial xi$. For n=1 both formulas give the same result, namely $X / \partial Y / \partial x \partial / \partial x$, therefore $R_1^0 (G_1) = R_1^0 (G_2)$. For n2 one clearly has $R_1^0 (G_1) \ne R_1^0 (G_2)$.

THE DIFFERENTIAL

In this section we express the restriction of the Chevalley–Eilenberg differential onto the sub complex $C_{GL_n}^* (ngl_n^{(\infty)}, \text{Map}(\mathcal{F}^{(\infty)} \mathcal{G})$ of GL_n-equivariant cochains in terms of graph complexes. Let us describe first the bracket in the limit $ngl_n^{(\infty)} = \bigcup_{s\ge 2} ngl_n^{(S)}$ of Lie algebras $ngl_n^{(S)}$ recalled in (8). If finite sums $a=a_2+a_3+a_4+\cdots$ and $b=b_2+b_3+b_4+\cdots$ are elements of $ngl_n^{(\infty)}$, $a_u, b_u \in \text{Sym}((\mathbb{R}^n)^{\otimes u}, \mathbb{R}^n)$, u2, then $[a,b]=[a,b]_3+[a,b]_4+\cdots$ (no quadratic term) with

$$[a,b]_u = \sum_{s+t=u+1} \sum_{1\leqslant i\leqslant s} \left(S(a_s \circ_i b_t) - S(b_s \circ_i a_t) \right)$$

where $S(-)$ denotes the symmetrization (see Remark 4.5) of a linear map $\mathbb{R}^{n\otimes u} \to \mathbb{R}^n$, $a_s \circ_i b_t$ is the insertion of b_t into the i th slot of a_s and $b_s \circ_i a_t$ has the similar obvious meaning. For $v_1,\ldots,v_u \in \mathbb{R}_n$ we easily get

$$[a,b]_u(v_1,\ldots,v_u) = \left| \sum_{s+t=u+1} \frac{s!t!}{u!} \sum_{\sigma} \{ a_s \left(b_t(v_{\sigma(1)},\ldots,v_{\sigma(t)}), v_{\sigma(t+1)},\ldots,v_{\sigma(u)} \right) \right.$$

$$\left. - b_s \left(a_t(v_{\sigma(1)},\ldots,v_{\sigma(t)}), v_{\sigma(t+1)},\ldots,v_{\sigma(u)} \right) \right\}, \tag{28}$$

where σ runs over all $(t,s-1)$-unshuffles σ, i.e. permutations $\sigma \in \Sigma_u$ such that $\sigma(1)<\cdots<\sigma(t),\sigma(t+1)<\cdots<\sigma(u)$.

Remark 4.1: In the rest of the paper, we will consider $\mathrm{ngl}_n^{(\infty)}$ with the modified Lie bracket, given by formula (28) without the $\dfrac{S!t!}{u!}$-coefficients. Since this modified Lie algebra is isomorphic to the original one, via the isomorphism $a_s \mapsto s!\cdot a_{s'}$ for $a_s \in \mathrm{Sym}((\mathbb{R}^n)^{\otimes s}, \mathbb{R}^n)$, $S\geq 2$, our modification is purely conventional. The advantage of this modified bracket is that the corresponding replacement rule (29) is a linear combination of graphs without fractional coefficients.

To help the reader to appreciate the idea of the differential, we start with an informal definition. A precise formula including signs and orientations is given in (32). At the beginning of Section 2 we decomposed the CE-differential into the sum $\delta_{CE}=\delta_1+\delta_2$. Let us analyze the action of the second piece δ_2 first. A graph G representing a GL_n-invariant m-cochain has m white vertices that mark the places where to insert elements of $\mathrm{ngl}_n^{(\infty)}$. Let us label, for m1, these white vertices by $\ell \in \{1,\ldots,m\}$ and denote the vertex labelled ℓ by $w\ell$. If m=0, there are no white vertices and no labelling is necessary.

The effect of the differential δ_2 on the graph G is, by the definition recalled in (5), the following. For each $\ell \in \{1,\ldots,m\}$ insert to the vertex wℓ the element $[h_i, h_j]$ and to the remaining white vertices elements $h_1, \cdots, \hat{h}_i, \cdots, \hat{h}_j, \cdots, h_{m+1}$, make the summation over all $1 \leq i < j \leq m+1$ and antisymmetrize in $h_1, \ldots, h_{m+1}$. Denote the resulting (m+1)-cochain by G_ℓ. Then $\delta_2(G) = \varepsilon_1 \cdot G_1 + \cdots + \varepsilon_1 \cdot G_{m'}$, where $\varepsilon 1, \ldots, \varepsilon_m \in \{-1, +1\}$ are appropriate signs. A moment's reflection reveals that G_ℓ is obtained by replacing the vertex w_ℓ by:

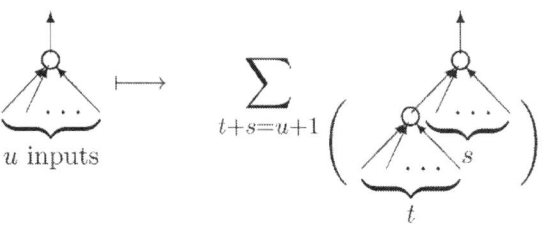

(29)

where the braces $(-)_{ush}$ indicate that the summation over all $(t, s-1)$-unshuffles of the inputs has been performed. This is precisely the formula for the generators of the homological vector field introduced by Merkulov [19] and [21]. One also recognizes (29) as the graphical representation of the axioms of L_∞-algebras as given in [12, page 160].

A similar analysis shows that $\delta 1$ acts by replacing each vertex of type 1 or 2 by the pictorial representation of the action of $ngl_n^{(\infty)}$ on tensors corresponding to this vertex. We will show instances of these 'pictorial presentations' in the following two examples.

Example 4.2

Consider a symmetric map $\xi: \mathbb{R}^{n \otimes v} \to \mathbb{R}^n$ representing an element in the fiber of the k th tangent space $T^{(k)}$ with coordinates $X^a_{(S_1, \cdots, S_v)}$ (see (16) of Example 3.2). The action of $a = a_2 + a_3 + a_4 \cdots \in ngr_n^{(\infty)}$ on ξ is given by $a\xi = (a\xi)_{u+1} + (a\xi)_{u+2} + \cdots$, where

$$(a\xi)_v = \sum_{s+u=v+1} \left(\sum_{1 \leq i \leq s} S(a_s \circ_i \xi) - \sum_{1 \leq i \leq v} S(\xi \circ_i a_s) \right)$$

.

Removing fractional coefficients by modifying the $\mathrm{ngl}_n^{(\infty)}$-action (compare Remark 4.1), one can graphically express the above rule by the following polarization of (29):

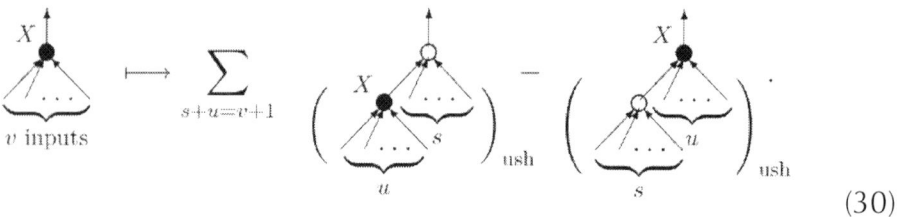

$$(30)$$

Example 4.3

Let us write two initial replacement rules for the connection and its derivatives. The first one is the infinitesimal version of (3):

The next one is a graphical form of an equation that can be found in [8, Section 17.7] (but notice a different convention for covariant derivatives used in [8]):

We are not going to give a general formula. For our purposes, it will be enough to know that it is of the form

$$(31)$$

where G_w is a linear combination of 2-vertex trees with one vertex (18), with v<w, and one vertex (19) with u<w+2.

Let us write a formal definition of the graph differential. For each oriented graph $G \in \mathcal{G}r_{\mathfrak{F},\mathfrak{G}}^m$

we define $\delta(G) \in \mathcal{G}r_{\mathfrak{F},\mathfrak{G}}^{m+1}$ as the sum over the set $\mathrm{Vert}(G)$ of vertices of G,

$$\delta(G) = \sum_{v \in \mathit{Vert}(G)} \varepsilon_v \cdot \delta_v(G)$$

$$(32)$$

where δ_v is the replacement of the vertex v determined by the type of v and geometric data as explained above. The signs ε_v and the orientations of the graphs in $\delta_v(G)$ are determined in the following way.

i. The operation δ_v replaces a 1st or 2nd type vertex v by a linear combination of graphs containing precisely one white vertex. The orientation of the graphs in $\delta_v(G)$ is given by the unique linear order such that this new white vertex is the minimal element and the relative order of the remaining white vertices is unchanged. The sign ε_v is +1. Symbolically

$$\delta_v(\circ < \cdots < \circ) = +1 \cdot \left(\delta_v(\bullet) < \circ < \cdots < \circ \right)$$

$$(33).$$

ii. (ii) Let v be a white vertex. We may assume that, after changing the sign of the graph G if necessary, v is the minimal element in an order determining the orientation. The orientation of graphs in $\delta_v(G)$ is then given by requiring that the lower left white vertex in the right-hand side of (29) is the minimal one, the upper right white vertex of (29) is the next one, and that the relative order of the remaining white vertices is unchanged. The sign ε_v is again +1. Symbolically,

$$\delta_v(\circ < \cdots < \circ) = +1 \cdot \left(\delta_v(\circ) < \circ < \cdots < \circ \right).$$ We leave as a simple exercise to derive from the rule (ii) that, if the white vertex v is the ith element of a linear order determining the orientation of G, for some $1 \leqslant i \leqslant m$, the orientations of graphs in $\delta_v(G)$ are symbolically expressed as

$$\delta_V(\circ < \cdots < \circ) = (-1)^{i+1} \cdot \left(\underbrace{\circ < \cdots < \circ}_{i-1} < \delta_V(\circ) < \underbrace{\circ < \cdots < \circ}_{m-i} \right)$$

(34)

Let us emphasize that the applications in this paper use only the initial part $\delta : \mathcal{G}r^0_{\mathcal{F},\mathcal{G}} \to \mathcal{G}r^1_{\mathcal{F},\mathcal{G}}$ of the differential. Since the graphs spanning $\mathcal{G}r^0_{\mathcal{F},\mathcal{G}}$ (resp. $\mathcal{G}r^1_{\mathcal{F},\mathcal{G}}$) have no white vertices (resp. one white vertex), the orientation issue is trivial and all ε_v's in (32) are +1.

Theorem 4.4: The object $\mathcal{G}r^*_{\mathcal{F},\mathcal{G}} = (\mathcal{G}r^*_{\mathcal{F},\mathcal{G}}, \delta)$, is a cochain complex and the maps R^m_n in (26)assemble into a cochain map

$$(\mathcal{G}r^*_{\mathcal{F},\mathcal{G}}, \delta) \to (C^*_{GL_n}(\mathrm{ngl}^{(\infty)}_n, \mathrm{Map}(\mathcal{F}^{(\infty)}, \mathcal{G})), \delta_{CE})$$

.

Proof: Using the antisymmetry of f, one can rewrite Eqs. (4) and (5) into

$$(\delta_1 f)(h_1, \ldots, h_{m+1}) = \frac{1}{m!} \mathrm{Ant}(h_1 f(h_2, \ldots, h_{m+1}))$$

$$(\delta_2 f)(h_1, \ldots, h_{m+1}) = \frac{1}{2(m-1)!} \mathrm{Ant}(f([h_2, h_1], h_3, \ldots, h_{m+1})),$$

where $\mathrm{Ant}(-)$ denotes the antisymmetrization, see Remark 4.5 below. If the multilinear map f itself is an antisymmetrization $\mathrm{Ant}(F)$ of a map F, one can rewrite the above displays into

$$(\delta_1 f)(h_1, \ldots, h_{m+1}) = \mathrm{Ant}(h_1 F(h_2, \ldots, h_{m+1}))$$

(35)

$$(\delta_2 f)(h_1, \ldots, h_{m+1}) = \mathrm{Ant}\left(\sum_{1 \leqslant i \leqslant m} (-1)^{i+1} F(h_1, \ldots, h_{i-1}, [h_{i+1}, h_i], h_{i+2}, \ldots, h_{m+1}) \right)$$

(36)

After this preparation, we prove that R^*_n is a chain map by verifying that $(R^{m+1}_n \circ \delta)(G) = (\delta_{CE} \circ R^m_n)(G)$ for each graph G generating $\mathcal{G}r^m_{\mathcal{F},\mathcal{G}}$. After choosing a linear order of white vertices of G compatible with its orientation, an appropriate version of the 'state sum' (11) of [11] gives a multilinear map F such that $R^m_n(G) = \mathrm{Ant}(F)$, see [11, Remark 5.2].

It is not difficult to see that R_n^* translates the part of the differential $\delta(G)$ in (32) given by the summation over the 1st and 2nd type vertices into formula (35) for $\delta_1(f)$ and the part of $\delta(G)$ given by the summation over the white vertices to formula (36) for $\delta_2(f)$. This fact is also reflected by the obvious similarity between formulas (35) and (36) for the Chevalley–Eilenberg differential and symbolic formulas (33) and (34) for the graph differential.

The condition $\delta^2=0$ can be verified directly using the fact that the local replacement rules used in (32) are duals of Lie algebra actions and checking that the orientations were defined in such a way that the signs combine properly. One may, however, proceed also as follows.

Since both the domain and target of the map R_n^*, as well as R_n^* itself, are defined in terms of "standard representations", R_n^m makes sense for an arbitrary natural n. Let $G \in \mathcal{G}r_{\mathfrak{F},\mathfrak{G}}^m$. By the finitary nature of objects involved, there exists e0 such that all graphs that constitute $\delta^2(G) \in \mathcal{G}r_{\mathfrak{F},\mathfrak{G}}^{m+2}$ have e edges. Choose ne−m−2. We already know that R_n^* commutes with the differentials, therefore $R_n^{m+2}\left(\delta^2(G)\right) = \delta_{CE}^2\left(R_n^m(G)\right) = 0$. By the second part of Theorem 3.3 this implies that $\delta^2(G)=0$.

Remark 4.5: In this paper, the antisymmetrization of an element x of some (say) right Σ_k-module, k1, is given by the formula $\text{Ant}(x):= \sum_{\sigma \in \Sigma_k} \text{sgn}(\sigma) \cdot x\sigma$, without the traditional $\frac{1}{k!}$. This convention is forced by the standard definition of the Lie algebra associated to an associative algebra $(A,\cdot)$—the bracket $[a',a"]:=a'\cdot a"−a"\cdot a'$, $a',a"\in A$, does not involve the $\frac{1}{2!}$-factor. On the other hand, we define the symmetrization of x as above by the expected formula $S(x) := \frac{1}{k!}\sum_{\sigma \in \Sigma_k} x\sigma$.

Remark 4.6: Applications of our theory will often be based on a suitable choice of a subspace of $\mathfrak{Nat}(\mathfrak{F},\mathfrak{G})$, together with the corresponding subcomplex of the graph complex $\mathcal{G}r_{\mathfrak{F},\mathfrak{G}}^*$. These subobjects, denoted for the purposes of this remark by $\mathfrak{Nat}(\mathfrak{F},\mathfrak{G})$ and $\mathcal{G}r_{-\mathfrak{F},\mathfrak{G}}^* = (\mathcal{G}r_{-\mathfrak{F},\mathfrak{G}}^*,\delta)$, will be chosen so that the number of edges of graphs spanning $\mathcal{G}r^*_{-\mathfrak{F},\mathfrak{G}}$ will be, for each $m \geqslant 0$, bounded by C+m, where C is a fixed constant.

An example is the subcomplex $\mathcal{G}r^*_\bullet(d)$ of the graph complex $\mathcal{G}r^*_{T\times d,T}$, introduced in Section 5, that describes d -multilinear operators on vector fields. Graphs spanning $\mathcal{G}r^*_\bullet(d)$ have precisely d+m edges, so C=d for this subcomplex. Another example is the subcomplex

$\mathcal{G}r^*_{\bullet\Lambda}(d)$ of $\mathcal{G}r^*_{\text{Conx}T\times d,T}$ describing 'connected' d-multilinear operators used in Section 7. Each degree m graph spanning this subcomplex has at most 2d+m−1 edges, i.e. C=2d−1 in this case. The third example is the complex $\mathcal{G}r^*_{\bullet\Lambda_{\circlearrowright}}(d)$ introduced on page 270 describing 'connected' operators in $\mathfrak{Nat}(\text{conx } T^{\otimes d}, \mathfrak{R})$. For this complex, C:=2d.

Let $\left(\mathcal{G}r^*_{-\mathfrak{F},\mathfrak{G}}, \underline{\delta}\right)$, $\mathfrak{Nat}(\mathfrak{F},\mathfrak{G})$ and the constant C be as above. By Theorem 2.2 combined with Theorem 4.4, the restriction R^*_{-n} of R^*_n induces the map

$$H^0(\underline{R}^*_n) : H^0(\mathcal{G}r^*_{-\mathfrak{F},\mathfrak{G}}, \underline{\delta}) \rightarrow \mathfrak{Nat}(\mathfrak{F}, \mathfrak{G})$$

$$(37)$$

which is an isomorphism in stable dimensions. By this we mean that the dimension n of the underlying manifold M is C. If this happens, then the map $\underline{R}^*_n$ is, by [11, Proposition 4.9], a chain isomorphism, so $H^0\left(\underline{R}^*_m\right)$ is an isomorphism, too. If the dimension of M is less than the stable dimension, one cannot say anything about the induced map $H^0\left(\underline{R}^*_m\right)$, although the chain map $\underline{R}^*_n$ is still a chain epimorphism.

Example 4.7
In this example we prove a baby version of Theorem 5.1. Namely, we show that the only natural bilinear operations on vector fields on manifolds of dimensions $\geqslant 2$ are scalar multiples of the Lie bracket. It will be convenient to have ready some initial cases of formula (30) for the replacement rule of vertices representing vector fields and their derivatives:

$$\delta\left(\begin{array}{c}\uparrow\\\bullet\end{array}\right)=0,\ \delta\left(\begin{array}{c}\uparrow\\\bullet\end{array}\right)=\ \ ,\ \ \delta\left(\ \begin{array}{c}\end{array}\ \right)=-\ \ +\ \ +\ \ ,\ \ \ldots$$

It is also clear that $\delta\left(\overset{\square}{\uparrow}\right)=0$.

Let us denote by $\mathcal{Gr}^{*}_{T\otimes T,T}\subset\mathcal{Gr}^{*}_{T\times T,T}$ the subcomplex describing bilinear operators. Its degree 0 part $\mathcal{Gr}^{0}_{T\otimes T,T}$ is spanned by

$$\begin{array}{c}\blacksquare\\\bullet\,Y\\\bullet\,X\end{array}\ ,\quad\begin{array}{c}\blacksquare\\\bullet\,X\\\bullet\,Y\end{array}\ ,\quad\begin{array}{c}\blacksquare\ Y\\\bullet\,X\ \bigcirc\end{array}\quad\text{and}\quad\begin{array}{c}\blacksquare\ X\\\bullet\,Y\ \bigcirc\end{array}$$

One easily calculates the differential of the leftmost term:

$$\delta\left(\begin{array}{c}\blacksquare\\\bullet\,Y\\\bullet\,X\end{array}\right)=\begin{array}{c}\delta(\blacksquare)\\\bullet\,Y\\\bullet\,X\end{array}+\delta\left(\begin{array}{c}\blacksquare\\\bullet\,Y\end{array}\right)+\begin{array}{c}\blacksquare\\\bullet\,Y\\\delta(\bullet\,X)\end{array}=\begin{array}{c}\blacksquare\\\ Y\\\bullet\,X\end{array}\in\mathcal{Gr}^{1}_{T\otimes T,T}$$

$$(38)$$

and similarly one gets

$$\delta\left(\begin{array}{c}\blacksquare\ Y\\\bullet\,X\ \bigcirc\end{array}\right)=\begin{array}{c}\blacksquare\\\bullet\,X\ \ Y\ \bigcirc\end{array}\in\mathcal{Gr}^{1}_{T\otimes T,T}$$

The formula for the differential of the remaining two generators of $\mathcal{Gr}^{0}_{T\otimes T,T}$ is obtained by interchanging $X\leftrightarrow Y$ in the previous two displays. One clearly has

$$\delta\left(\begin{array}{c}\blacksquare\ \ \ \ \blacksquare\\\bullet\,Y\text{---}\bullet\,X\\\bullet\,X\ \ \ \ \bullet\,Y\end{array}\right)=\begin{array}{c}\ Y\\\bullet\,X\end{array}-\begin{array}{c}\ X\\\bullet\,Y\end{array}=0$$

because the inputs of white vertices are symmetric. It is easy to verify that the element

$$b := \quad \begin{array}{c} \bullet\, Y \\ \bullet\, X \end{array} - \begin{array}{c} \bullet\, X \\ \bullet\, Y \end{array} \in \mathcal{Gr}^0_{T \otimes T, T}$$

(39)

representing the Lie bracket in fact spans all cochains in $\mathcal{Gr}^0_{T \otimes T, T}$. We conclude that $H^0\left(\mathcal{Gr}^0_{T \otimes T, T}, \delta\right)$ is one-dimensional, generated by the bracket [X,Y]. The complex $\mathcal{Gr}^0_{T \otimes T, T}$ clearly fits into the scheme discussed in Remark 4.6 (with C=2), which proves Theorem 5.1 for d=2.

Example 4.8

We close this section by an example suggested by the referee which will further illuminate the meaning of the graph differential. The graph

$$\begin{array}{c} \blacksquare \\ \uparrow \\ \bullet\, Y \\ \bullet\, X \end{array}$$

(40)

represents the local expression

$$\left(X^i \frac{\partial}{\partial x^i},\ Y^i \frac{\partial}{\partial x^i} \right) \longmapsto X^i \frac{\partial Y^j}{\partial x^i} \frac{\partial}{\partial x^j}$$

(41)

If $\{y^i\}$ is a different set of coordinates, then X and Y transforms to $X^i \frac{\partial y^r}{\partial x^j} \frac{\partial}{\partial y^s}$ and $Y^j \frac{\partial y^r}{\partial x^j} \frac{\partial}{\partial y^r}$, respectively. Having this transformed X act on the transformed Y gives

$$X^i \frac{\partial y^s}{\partial x^i} \frac{\partial}{\partial y^s} \left(Y^j \frac{\partial y^r}{\partial x^j} \right) \frac{\partial}{\partial y^r} = X^i \frac{\partial y^s}{\partial x^i} \frac{\partial Y^j}{\partial y^s} \frac{\partial y^r}{\partial x^j} \frac{\partial}{\partial y^r} + X^i \frac{\partial y^s}{\partial x^i} Y^j \frac{\partial}{\partial y^s} \left(\frac{\partial y^r}{\partial x^j} \right) \frac{\partial}{\partial y^r}$$

The first term in the right-hand side is equal to the expression in (41) under change-of-coordinates, so the second term represents the extent to which this expression is not invariant. It is equal to $X^i Y^j \partial^2 y^r / \partial x^i \partial x^j \, \partial / \partial y^r$, which translates directly to the formula (38) for the differential of (40) in the graph complex.

OPERATIONS ON VECTOR FIELDS

In this section we consider differential operators acting on a finite number of vector fields X,Y,Z,... with values in vector fields, that is, operators in $\mathfrak{Nat}(T^{\times \infty}, T) = U_{d \geq 0} \mathfrak{Nat}(T^{\times d}, T)$. The first statement of this section is:

Theorem 5.1
Let M be a smooth manifold and d a natural number such that dim(M) $\geq$ d. Then each d-multilinear natural operator from vector fields to vector fields is a sum of iterations of the Lie bracket containing each of d variables precisely once, and all relations between these expressions follow from the Jacobi identity and antisymmetry. In particular, there are precisely (d−1)! linearly independent operators of the above type.

Theorem 5.1 is an obvious consequence of Proposition 5.6 below and the formula for the dimension of the k th piece of the operad $\mathfrak{Lie}$ for Lie algebras that can be found for example in [6, Example 3.1.12]. Theorem 5.1 describes multilinear operators and does not cover operators as $\mathfrak{O}$ (X,Y,Z):=[X,Y]+[X,[X,Z]] but can easily be extended to cover also these cases. Since all operators are assumed to be polynomial, they decompose into the sum of their homogeneous parts. For instance, $\mathfrak{O}$ (X,Y,Z) is the sum of the homogeneity-2 part [X,Y] and the homogeneity-3 part [X,[X,Z]].

Remark 5.2

Let us explain the decomposition of operators $\mathfrak{O} \in \mathfrak{Nat}(T^{\times \infty}, T)$ into homogeneous parts in more detail. The local formula O for the operator $\mathfrak{O}$ is the sum $O = O_1 + \cdots + O_r$, where O_d is the part of O consisting of terms with precisely d occurrences of the vector field variables. The action of the structure group $GL_n^{(\infty)}$ on the typical fiber of the prolongation of $T^{\times\infty}$ is linear, which is expressed by the manifest linearity of the replacement rule (30) in the vector field variable. This implies that the map O is $-GL_n^{(\infty)}$ equivariant if and only if each of its homogeneous components Od is $GL_n^{(\infty)}$-equivariant, 1dr. Therefore $\mathfrak{O} = \mathfrak{O}_1 + \cdots + \mathfrak{O}_{r'}$, where $\mathfrak{O}_d$ is the operator defined by the local formula $O_{d'}$, 1dr.

We conclude that to classify operators of the above type, it suffices to classify homogeneous operators. It is a standard fact that each homogeneous operator of degree d is either d-multilinear or a sum of operators obtained from d-multilinear operators by repeating one or more of their variables. We will call this procedure the depolarization of multilinear operators. Theorem 5.1 therefore implies the following corollary.

Corollary 5.3

Let M be a smooth manifold. Each natural differential operator from vector fields on M to vector fields on M whose all components are of homogeneity $\leq \dim(M)$is a sum of iterations of the Lie bracket. All relations between these iterations follow from the Jacobi identity and antisymmetry.

In Example 4.7 we studied the graph complex $\mathcal{Gr}^*_{T\otimes T,T}$ describing bilinear operators. Bearing this example in mind, we introduce $\mathcal{Gr}^*_\bullet(d) = \mathcal{Gr}^*_{T\otimes T,T} \subset \mathcal{Gr}^*_{T^{\times d},T}$, the sub complex describing d-multilinker operators. Its degree m component is spanned by graphs with d vertices of the first type labelled by $X_1, \ldots, X_d$, m white vertices of the third type and one 2nd type vertex $\dagger$ which we call the anchor . Observe that $\mathcal{Gr}^*_\bullet(d)$ is precisely the graph complex $\mathcal{Gr}^*_{\bullet(b)\Delta(C)}$ of [11, Corollary 5.1] with b:=d and c:=0. The collection $\mathcal{Gr}^0_\bullet = \{\mathcal{Gr}^0_\bullet(d)\}_{d\geq 1}$ of degree 0 subspaces admits two types of operations.

i. For graphs $G' \varepsilon \mathcal{Gr}^0_\bullet(u)$, $G'' \varepsilon \mathcal{Gr}^0_\bullet(v)$ and 1iu, one has the $\circ_i$-product $G' \circ i G'' \varepsilon \mathcal{Gr}^0_\bullet(u+v-1)$ given by the following straightforward extension of the Chapoton–Livernet vertex insertion [1, § 1.5] to non-simply connected graphs. Assume that $x'_1, \cdots, x'_u$ are the black vertices of G', $X''_1, \cdots, X''_v$ the black vertices of G'' and $\mathrm{In}(X'_i)$ the set of inputs of X' in G'. Then

$$G' \circ_i G'' := \sum_{f:In(X'_i) \to \{X''_1, \ldots, X''_v\}} G' \circ_i^f G'' \in \mathcal{G}r_\bullet^0(u + v - 1)$$

where $G' \circ_i^f G'' \in \mathcal{G}r_\bullet^0(u+v-1)$ is the graph obtained by replacing the vertex X'_i of G' by G'' and grafting the inputs of X'_i on black vertices of G'' following f.

In more detail, one starts by cutting off the anchor ⌐ of G'' and grafts the resulting free edge on the vertex of G' immediately above X'_i. Then one grafts each input edge e of X'_i on the vertex $f(e)$ of G''. Finally, one changes the labels $X'_i, \cdots, X'_{i-1}, X''_1, \cdots, X''_v, X'_{i+1}, \cdots, X'_u$ of the black vertices of the graph obtained in this way into $X_1, \ldots, X_{u+v-1}$.

ii. One has the right action of the symmetric group: for each $\mathcal{G} \in \mathcal{G}r_\bullet^0(d)$ and a permutation $\sigma \in \Sigma_d$, one has $\mathcal{G}\sigma \in \mathcal{G}r_\bullet^0(d)$ given by permuting the labels $X_1, \ldots, X_d$ of the black vertices of G according to σ.

Proposition 5.4

The collection $\mathcal{G}r_\bullet^0 = \{\mathcal{G}r_\bullet^0(d)\}_{d \geq 1}$ with the above operations is an operad with unit $\mathbb{1} \in \mathcal{G}r_\bullet^0(1)$ [16]. The operad structure of $\mathcal{G}r_\bullet^0$ restricts to $H^0(\mathcal{G}r_\bullet^*, \delta) = \mathrm{Ker}(\delta : \mathcal{G}r_\bullet^0 \to \mathcal{G}r_\bullet^1)$.

Proof: The operad axioms for the operations in (i) and (ii) above are verified directly, compare also [1, § 1.5]. The simplest way to see that the operad structure of $\mathcal{G}r_\bullet^0$ restricts to the kernel of δ is to extend the operations (i) and (ii), in the obvious manner, to the graded collection $\mathcal{G}r_\bullet^*$, making $(\mathcal{G}r_\bullet^*, \delta)$ a dg-operad. This, in particular, would mean that δ is a derivation with respect to these extended $\circ_i$-operations, which implies the second part of the proposition.

Example 5.5

An instructive example of the vertex insertion can be found in [1, § 1.5]. We present here a simpler one, taken from the proof of [1, Theorem 1.9]. Let p be the graph

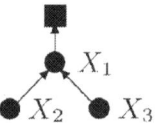

$$X_1 \in \mathcal{Gr}^0_\bullet(2)$$

Then one has

$$\mathbf{p} \circ_1 \mathbf{p} = \begin{array}{c}X_1 \\ X_2 \\ X_3\end{array} + X_1 \quad X_3 \in \mathcal{Gr}^0_\bullet(3) \quad \text{and} \quad \mathbf{p} \circ_2 \mathbf{p} = \begin{array}{c}X_1 \\ X_2 \\ X_3\end{array} \in \mathcal{Gr}^0_\bullet(3)$$

The above display implies that the associator Ass $(p):=p\circ_1 p - p\circ_2 p$ equals

$$\begin{array}{c}X_1 \\ X_2 \quad X_3\end{array}$$

and is therefore symmetric in X_2 and X_3. This, by definition, means that p represents a pre-Lie multiplication [1, § 1.1]. We will see that $\mathcal{Gr}^0_\bullet$ is indeed closely related to the pre-Lie operad pLie.

Let $\tau \in \Sigma_2$ be the generator. By standard properties of pre-Lie algebras [1, Proposition 1.2], the antisymmetrization $p(\tau-1)$ of the element p from Example 5.5 is a Lie bracket. Observe that $p(\tau-1)$ equals the element b introduced in (39).

Proposition 5.6

The 0th cohomology $H^0\left(\mathcal{Gr}^*_\bullet(d),\delta\right)$ is, for each $d\geq 2$, generated by the Lie bracket $b = p(\tau-1) \in H^0\left(\mathcal{Gr}^*_\bullet(2),\delta\right)$, by iterating operations (i) and (ii) above. There are no relations between these iterations other than those following from the Jacobi identity and antisymmetry.

$$\delta\left(\begin{array}{c}\bullet\\:\end{array}\right) = \begin{array}{c}\\:\end{array} + \text{ replacements of remaining vertices of the graph,}$$

$$\delta\left(\begin{array}{c}\bullet\\:\end{array}\right) = \begin{array}{c}\\:\end{array} + \begin{array}{c}\\:\end{array} + \begin{array}{c}\\:\end{array} - \begin{array}{c}\\:\end{array} + \cdots$$

$$\delta\left(\begin{array}{c}\bullet\\:\end{array}\right) = \begin{array}{c}\\:\end{array} + \begin{array}{c}\\:\end{array} + \begin{array}{c}\\:\end{array} + \begin{array}{c}\\:\end{array} - \begin{array}{c}\\:\end{array} + \cdots$$

Figure 1: Action of δ on $\mathcal{G}r^0_{\bullet\circ}$ —the replacement rule for a type 1 vertex on the wheel.

A compact formulation of Proposition 5.6 is that the operad $H^0\left(\mathcal{G}r^*_\bullet, \delta\right) = \left\{H^0\left(\mathcal{G}r^*_\bullet(d), \delta\right)\right\}_{d\geq 1}$ is isomorphic to the operad $\mathcal{L}ie = \{\mathcal{L}ie(d)\}_{d\geq 1}$ for Lie algebras [16, Example II.3.34], via an isomorphism that sends the generator $\beta \in \mathcal{L}ie(2)$ of $\mathcal{L}ie$ into $b \in \mathcal{G}r^0_\bullet(2)$. Graphs spanning $\mathcal{G}r^0_\bullet(d)$ have d edges which explains the stability condition $\dim(M) \geq d$ in Theorem 5.1. The rest of this section is devoted to a proof of its main result.

Proof of Proposition 5.6

It is clear from formulas (29) and (30) and $\delta\left(\begin{array}{c}\\\end{array}\right) = 0$ that the differential preserves connected components of underlying graphs. Therefore, for each $d \geq 1$, $\mathcal{G}r^*_\bullet(d)$ is the direct sum $\mathcal{G}r^*_\bullet(d) = \oplus_{c\geq 1}\mathcal{G}r^*_{\bullet c}(d)$, where $\mathcal{G}r^*_{\bullet c}(d)$ denotes the subcomplex spanned by graphs with c connected components. In particular, $\mathcal{G}r^*_{\bullet 1}(d)$ is the subcomplex of connected graphs. It is easy to see that $\mathcal{G}r^0_{\bullet 1}$ is a suboperad of $\mathcal{G}r^*_\bullet$.

As the Lie bracket represented by $b \in \mathcal{G}r^0_{\bullet 1}(2)$ is antisymmetric and satisfies the Jacobi identity, the rule $F(\beta) := b$, where $\beta \in \mathcal{L}ie(2)$ is the generator, defines an operad homomorphism $F : \mathcal{L}ie \to \mathcal{G}r^0_{\bullet 1}$. Since the Lie bracket and its iterations are natural operators, $\mathfrak{F}(F) \subset \mathrm{Ker}\left(\delta : \mathcal{G}r^0_{\bullet 1} \to \mathcal{G}r^1_{\bullet 1}\right)$. Proposition 5.6 will clearly be established if we prove that

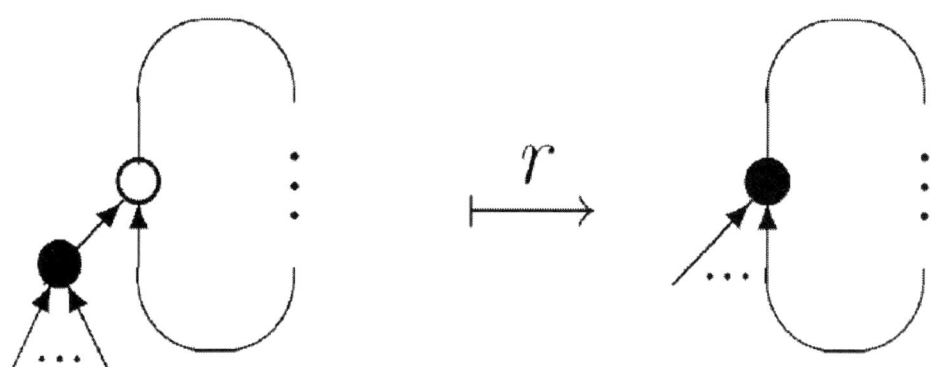

Figure 2: The map $r : \overline{\mathcal{G}r}^1_{\bullet\circlearrowleft}(d,w) \to \mathcal{G}r^0_{\bullet\circlearrowleft}(d,w)$ contracts the unique edge connecting the binary white vertex on the wheel with a black vertex outside the wheel.

i. the operad map $F : \mathcal{L}ie \to \mathcal{G}r^0_{\bullet 1}$ induces an isomorphism $\mathcal{L}ie \cong \mathcal{G}r^0_{\bullet 1} H^0\left(\mathcal{G}r^*_{\bullet 1}, \delta\right)$, and

ii. $H^0\left(\mathcal{G}r^*_{\bullet c}(d), \delta\right) = 0$, for each $c \geq 2$, $d \geq 1$.

Part (i) is highly nontrivial, but it in fact has already been proved in [14]. Indeed, the operad $\mathcal{G}r^0_{\bullet 1}$ is precisely the operad $p\mathcal{L}ie$ describing pre-Lie algebras [1] and $F : \mathcal{L}ie \to \mathcal{G}r^0_{\bullet 1}$ corresponds, under the identification $\mathcal{G}r^0_{\bullet 1} \cong p\mathcal{L}ie$, to the inclusion $\iota : \mathcal{L}ie \hookrightarrow p\mathcal{L}ie$ induced by the antisymmetrization of the pre-Lie product. The dg operad rpL∗ of [14] coincides, in degrees 0 and 1, with the complex $\mathcal{G}r^*_{\bullet 1}$ and the isomorphism in (i) is isomorphism (2) of [14].

Let us prove (ii). For each $m \geq 0$, $d \geq 1$, consider the span $\mathcal{G}r^m_{\bullet\circlearrowleft}(d)$ of connected graphs with d vertices $X_1, \ldots, X_d$ of type 1, m 'white' vertices of type 3 and no vertex of type 2. The direct sum $\mathcal{G}r^m_{\bullet\circlearrowleft}(d) \oplus_{m \geq 0} \mathcal{G}r^m_{\bullet\circlearrowleft}(d)$ is a cochain complex, with the differential defined in the same way as the differential in $\mathcal{G}r^*_{\bullet}(d)$ and denoted again by δ. We claim that, for each c2 and d1, there is an isomorphism of cochain complexes

$$\mathcal{Gr}^*_{\bullet_c}(d) \cong \bigoplus_{i_1+\cdots+i_c=d} \mathcal{Gr}^*_{\bullet_1}(i_1) \otimes \left(\mathcal{Gr}^*_{\bullet_\bigcirc}(i_2) \odot \cdots \odot \mathcal{Gr}^*_{\bullet_\bigcirc}(i_c) \right)$$

(42)

where $\odot$ as usual denotes the symmetric product. To prove this iso-morphism, observe that each graph $G \in \mathcal{Gr}^*_{\bullet_c}(d)$ decomposes into the disjoint union

$$G = G_1 \sqcup G_2 \sqcup \cdots \sqcup G_c,$$

(43)

of its connected components. Precisely one of these components contains the unique type 2 vertex $\blacksquare$, assume it is G1. Then $G_1 \in \mathcal{Gr}^*_{\bullet_1}(i_1)$ and $G_s \in \mathcal{Gr}^*_{\bullet_\bigcirc}(i_s)$ for $2 \le s \le c$, with some $i_1+\cdots+i_c=d$. Decomposition (43) is clearly unique up to the order of $G_2,\ldots,G_c$ and is preserved by the differential. This proves (42). By Künneth and Mashke's theorems, (ii) follows from $H^0\left(\mathcal{Gr}^*_{\bullet_\bigcirc}(d), \delta \right) = 0$, $d \ge 1$, which is the same as showing that

the map $\delta : \mathcal{Gr}^0_{\bullet_\bigcirc}(d) \to \mathcal{Gr}^1_{\bullet_\bigcirc}(d)$ is a monomorphism for each $d \ge 1$. (44)

Let us inspect the structure of $\mathcal{Gr}^*_{\bullet_\bigcirc}(d)$. It is clear from simple graph combinatorics that each graph in $\mathcal{Gr}^m_{\bullet_\bigcirc}(d)$ has genus 1, therefore it contains a unique wheel. Denote $\mathcal{Gr}^m_{\bullet_\bigcirc}(d,\omega) \subset \mathcal{Gr}^m_{\bullet_\bigcirc}(d)$ the sub-space spanned by graphs that have precisely w vertices (of either type) on the wheel, $w \ge 0$. It is obvious from (29) and (30) that $\delta\left(\mathcal{Gr}^m_{\bullet_\bigcirc}(d,\omega) \right) \subset \mathcal{Gr}^{m+1}_{\bullet_\bigcirc}(d,\omega) \oplus \mathcal{Gr}^{m+1}_{\bullet_\bigcirc}(d,\omega+1)$, for $d \ge 1$, $w \ge 0$; see also Fig. 1. Let us denote by δ_0 the component of δ that preserves the number of vertices on the wheel and $\delta 1$ the component that raises it by one. We claim that in order to prove (44), it is enough to verify that

the map $\delta^0 : \mathcal{Gr}^0_{\bullet_\bigcirc}(d) \to \mathcal{Gr}^1_{\bullet_\bigcirc}(d)$ is a monomorphism for each $d \ge 1$. (45)

The spaces $\mathcal{Gr}^m_{\bullet_\bigcirc}(d,p)$ form a bicomplex $\mathcal{Gr}^{*,*}_{\bullet_\bigcirc}(d), \delta$ with $\mathcal{Gr}^{p,q}_{\bullet_\bigcirc}(d) := \mathcal{Gr}^{p+q}_{\bullet_\bigcirc}(d,p)$ and δ the sum $\delta0+\delta1$, where $\delta^0 : \mathcal{Gr}^{*,*}_{\bullet_\bigcirc}(d) :\to \mathcal{Gr}^{*,*+1}_{\bullet_\bigcirc}(d)$ and $\delta^0 : \mathcal{Gr}^{*,*}_{\bullet_\bigcirc}(d) :\to \mathcal{Gr}^{*+1,*}_{\bullet_\bigcirc}(d)$ are defined above. Condition (45) then implies (44) via a standard spectral sequence argument. The only subtlety is that our bicomplex is not a first quadrant one, thus the convergence of the related spectral sequence has to be checked. We therefore decided to prove the implication (45) $\Rightarrow$ (44) by the following elementary calculation.

Suppose that (44) does not hold and let $x \in \mathcal{G}r^0_{\bullet\circlearrowleft}(d)$ be such that $\delta(x)=0$ while $x \neq 0$. There exists a decomposition $x=x_a+x_{a+1}+\cdots+x_{a+s}$ with $x_w \in \mathcal{G}r^0_{\bullet\circlearrowleft}(d,w)$ for $a \leq w \leq a+s$ in which $x_a \neq 0$. Since $\delta_0(x_a)$ is the component of $\delta(x)$ in $\mathcal{G}r^1_{\bullet\circlearrowleft}(d,a)$, $\delta_0(x_a)=0$. Then (45) implies $x_a=0$, a contradiction.

Denote by $\overline{\mathcal{G}}^1_{\bullet\circlearrowleft}(d,w) \subset \mathcal{G}r^1_{\bullet\circlearrowleft}(d,w)$ the subspace spanned by graphs with one binary white vertex on the wheel, as in the left graph in Fig. 2. Both $\overline{\mathcal{G}}^1_{\bullet\circlearrowleft}(d,w)$ and $\overline{\mathcal{G}}^1_{\bullet\circlearrowleft}(d,w)$ have canonical bases provided by isomorphism classes of graphs, therefore one has a canonical projection $x : \mathcal{G}r^1_{\bullet\circlearrowleft}(d,w) \rightarrow \overline{\mathcal{G}}^1_{\bullet\circlearrowleft}(d,w)$. In addition to the projection, there is a second map $r : \overline{\mathcal{G}}^1_{\bullet\circlearrowleft}(d,w) \rightarrow \mathcal{G}r^0_{\bullet\circlearrowleft}(d,w)$ whose definition is clear from Figure. 2.

Let $G \in \overline{\mathcal{G}}^1_{\bullet\circlearrowleft}(d,w)$ be a graph. Observe that $\mathcal{G}^0_{\bullet\circlearrowleft}(d,0)$, we may therefore assume $w \geq 1$. Recall that the differential $\delta(G)$ is the sum (32) of local replacements $\delta_v(G)$ over $v \in \text{Vert}(G)$. Let $\text{Vert}_\circlearrowleft(G) \subset \text{Vert}(G)$ be the subset of vertices on the wheel. For $v \in \text{Vert}_\circlearrowleft(G)$, the contribution $\delta v(G)$ contains precisely one graph in $\mathcal{G}r^1_{\bullet\circlearrowleft}(d,w)$ with the binary white vertex—see again Fig. 1. Denote this graph $\overline{\delta}^0(G)$ and define $\overline{\delta}^0(G) := \sum_{v \in \text{Ver}\circlearrowleft(G)} 0\overline{\delta}r^0_v(G)$.

It is clear that $\mathfrak{F}\left(\overline{\delta}r^0\right) \subset \overline{\mathcal{G}}^1_{\bullet\circlearrowleft}(d,w)$, $\overline{\delta}r^0 = \pi \circ \overline{\delta}^0$ and $r \circ \overline{\delta}^0 = \text{w.id}$. Combining these facts, we obtain $r \circ \pi \circ \delta^0 = w \cdot \text{id}$, which implies (45) and finishes the proof.

We believe that one can even show that the complex $\left(\mathcal{G}r^*_{\bullet\circlearrowleft}(d),\delta\right)$ used in the above proof is acyclic in all dimensions. Let us close this section by formulating the following interesting consequence of the proof of Proposition 5.6.

Corollary 5.7
In stable dimensions, there are no nontrivial differential operators from
vector fields to functions.

Proof: It is clear that d -multilinear operators from vector fields to functions are described by the graph complex $\mathcal{G}r^{*}_{\bullet_{\circ}}(d)$ introduced in our
proof of Proposition 5.6. Condition (44) implies that there are no non-
trivial d-multilinear operators of this type. The corollary then follows
from the standard (de)polarization trick.

STRUCTURE OF THE SPACE OF NATURAL OPERATORS

In Example 1.8 we considered the trivial natural bundle $\mathfrak{R}$ whose
sections are smooth functions. Let $\mathfrak{F}$ be another natural bundle. The
space $\mathfrak{Nat}(\mathfrak{F},\mathfrak{R})$ of natural operators $\mathfrak{O}:\mathfrak{F}\to\mathfrak{R}$ with the 'pointwise'
multiplication is a commutative algebra, with unit 1 the operator that
sends all sections of $\mathfrak{F}$ into the constant section $1\in\mathbb{R}$. This indicates
that spaces of natural operators may sometimes have a rich algebraic
structure that can be used to simplify their classification.

Definition 6.1
We say that $\mathfrak{F}$ is a bundle with connected replacement rules if the
replacement rules send a connected graph to a linear combination of
connected graphs.

All natural bundles considered in this paper have connected replacement rules, and the author does not know any 'natural' natural operator that has not. We will see that the space of natural operators
between bundles with connected replacement rules exhibits some
freeness property. Before we formulate the first statement of this type,
we introduce the following convention.

The graph complex $\mathcal{G}r^{*}_{\mathfrak{F},\mathfrak{R}}$ for operators in $\mathfrak{Nat}(\mathfrak{F},\mathfrak{R})$ is spanned by
graphs with vertices of the 1st type representing tensors in a prolonga-
tion of the fiber of $\mathfrak{F}$, vertices (18) of the third type and one 2nd type

vertex ■ which in this case has no inputs and no outputs. Therefore ■ is an isolated vertex bearing no information and we discard it from the picture. With this convention, graphs spanning $\mathcal{G}r^{*}_{\mathfrak{F},\mathfrak{R}}$ have vertices of the 1st and 3rd type only. The disjoint union of graphs spanning $\mathcal{G}r^{*}_{\mathfrak{F},\mathfrak{R}}$ translates into the pointwise multiplication of the corresponding operators and the unit $1 \in \mathfrak{Nat}(\mathfrak{F},\mathfrak{R})$ is represented by the 'exceptional' empty graph.

Theorem 6.2

Let $\mathfrak{F}$ be a natural bundle with connected replacement rules. Then, in stable dimensions, the commutative unital algebra $\mathfrak{Nat}(\mathfrak{F},\mathfrak{R})$ is free, generated by the subspace $\mathfrak{Nat}_1(\mathfrak{F},\mathfrak{R})$ of natural operators represented by connected graphs. In other words, $\mathfrak{Nat}(\mathfrak{F},\mathfrak{R}) \cong \mathbb{R}[\mathfrak{Nat}_1(\mathfrak{F},\mathfrak{R})]$, where $\mathbb{R}[-]$ denotes the polynomial algebra functor.

Proof: Each graph spanning $\mathcal{G}r^{*}_{\mathfrak{F},\mathfrak{R}}$ decomposes into the disjoint union of its connected components. The differential δ, by assumption, preserves this decomposition which is clearly unique up to the order of components. The proof is finished by recalling that the disjoint union of graphs expresses the pointwise multiplication of operators.

Let $\mathfrak{F}, \mathfrak{G}$ be natural bundles. The pointwise multiplication makes the space $\mathfrak{Nat}(\mathfrak{F},\mathfrak{G})$ a unital module over the unital algebra $\mathfrak{Nat}(\mathfrak{F},\mathfrak{R})$. We prove a structure theorem also for this space.

Theorem 6.3

Suppose that both $\mathfrak{F}$ and $\mathfrak{G}$ are bundles with connected replacement rules. Then, in stable dimensions, $\mathfrak{Nat}(\mathfrak{F},\mathfrak{G})$ is the free $\mathfrak{Nat}(\mathfrak{F},\mathfrak{R})$-module generated by the subspace $\mathfrak{Nat}_1(\mathfrak{F},\mathfrak{G})$ of operators represented by connected graphs,

$$\mathfrak{Nat}(\mathfrak{F}, \mathfrak{G}) \cong \mathfrak{Nat}_1(\mathfrak{F}, \mathfrak{G}) \otimes \mathfrak{Nat}(\mathfrak{F}, \mathfrak{R}).$$

(46)

Proof: The proof is similar to the proof of Theorem 6.2. The graph complex $\mathcal{Gr}^*_{\mathfrak{F},\mathfrak{R}}$ describing operators in $\mathfrak{Nat}(\mathfrak{F},\mathfrak{G})$ is spanned by graphs with vertices of the first and third types, and one vertex of the second type. Each such a graph is the disjoint union of its connected components as in (43) and the differential preserves this decomposition. Precisely one of these components contains the vertex of the third type thus representing an operator in $\mathfrak{Nat}_1(\mathfrak{F},\mathfrak{G})$. The remaining components describe operators from $\mathfrak{Nat}_1(\mathfrak{F},\mathfrak{R})$ and assemble, via the pointwise multiplication, into an operator in $\mathfrak{Nat}(\mathfrak{F},\mathfrak{R})$.

Theorem 6.2 and Theorem 6.3 imply that in order to classify operators in $\mathfrak{Nat}(\mathfrak{F},\mathfrak{G})$, it is enough to understand the 'connected' subspaces $\mathfrak{Nat}_1(\mathfrak{F},\mathfrak{R})$ and $\mathfrak{Nat}_1(\mathfrak{F},\mathfrak{G})$. We will use this fact in the next section.

Example 6.4
In Section 5 we studied natural operators on vector fields with values in vector fields, that is, operators in $\mathfrak{Nat}(T^{\times\infty},T):=U_{d\geq0}\mathfrak{Nat}(T^{\times d},T)$. We also considered operators with values in functions and proved, in Corollary 5.7, that there are no nontrivial operators of this type in stable dimensions.

This means that $\mathfrak{Nat}(T^{\times\infty},\mathfrak{R})$ is the trivial commutative algebra $\mathbb{R}$ and (46) reduces to the isomorphism $\mathfrak{Nat}(T^{\times\infty},T)\cong\mathfrak{Nat}_1(T^{\times\infty},T)$ which says that all operators from vector fields to vector fields live, in stable dimensions, on connected graphs.

OPERATORS ON CONNECTIONS AND VECTOR FIELDS

We will consider operators acting on a linear connection Γ and a finite number of vector fields $X,Y,Z,\ldots$, with values in vector fields, such as the covariant derivative $\nabla_X Y$, torsion $T(X,Y)$ and curvature $R(X,Y)$ Zrecalled in Example 1.2. By Theorem 6.2 and Theorem 6.3, the structure of the space $\mathfrak{Nat}(\mathrm{Con}\times T^{\times\infty},T):=U_{d\geq0}\mathfrak{Nat}(\mathrm{Con}\times T^{\times d},T)$ of these operators is determined by the 'connected' subspaces $\mathfrak{Nat}_1(\mathrm{Con}\times T^{\times\infty},T)$ and $\mathfrak{Nat}_1(\mathrm{Con}\times T^{\times\infty},\mathfrak{R})$. In this section we describe these spaces. The following remark should be compared to Remark 5.2 in Section 5.

Remark 7.1

The local formula O for a natural differential operator O in $\mathfrak{Nat}$ (Con×T$^{×∞}$,T) or in $\mathfrak{Nat}$ (Con×T$^{×∞}$,$\mathfrak{R}$) decomposes into $O=\sum_{a,b\geq 0}O_{a,b}$ (finite sum), where $O_{a,b}$ is the part of Ocontaining precisely a ∇-variables and b vector field variables. For example, the local formula (1) for the covariant derivative represented by the graph in (27) is the sum $O_{1,2}+O_{0,2}$, where $0_{1,2}(X,Y,\Gamma):=\Gamma^i_{jk}X^jY^k\partial/\partial x^i$ and $0_{0,2}(X,Y,\Gamma):=X^jY^i_j\partial/\partial x^i$.

In contrast to Section 5, here the action of the structure group $GL_n^{(\infty)}$ on the typical fiber is linear only in the vector-field variables—the non-linearity in the ∇-variables is manifested in the presence of the 'isolated' white vertex in the replacement rule (31). Nevertheless, one may still decompose $\mathfrak{D}=\mathfrak{D}_1+\cdots+\mathfrak{D}_{r'}$ with $\mathfrak{D}_k$ the operator represented by the local formula $O_d:=\sum_{a\geq 0}O_{a,d'}$ $1\leq d\leq r$. Therefore homogeneity andmultilinearity in this section always refer to the vector fields variables.

The first half of this section will be devoted to the study of the space Nat1(Con×T×∞,T), the space $\mathfrak{Nat}_1$(Con×T$^{×∞}$,$\mathfrak{R}$) will be addressed in the second half of this section. As in Section 5, we start with multilinear operators.

Theorem 7.2

Let d≥0. On smooth manifolds of dimension ≥2d−1, each d-multilinear operator in $\mathfrak{Nat}_1$(Con×T$^{\otimes d}$,T)is a linear combination of iterations of the covariant derivative and the Lie bracket which contains each of the vector fields $X_1,\ldots,X_d$ exactly once. All relations follow from the anti-commutativity and the Jacobi identity of the Lie bracket.

If g_d denotes the number of linearly independent operators of this type, the generating function $g(t)=\sum_{d\geq 1}\frac{1}{d}g_dt^d$ is determined by the functional equation

$$e^{g(t)}\left(1-t-g^2(t)\right)=1.$$

$$(47)$$

Eq. (47) can be expanded into inductive formula (53) from which one can calculate some initial values of g_k as $g_1=1$, $g_2=3$, $g_3=26$, &c. Theorem 7.2 will follow from Proposition 7.4 below. The depolarization of Theorem 7.2 is:

Corollary 7.3

On a smooth manifold M, each operator from $\mathfrak{Nat}_1(\mathrm{Con} \times T^{\times\infty}, T)$ whose all components are of homogeneity $\leq \frac{1}{2}(\dim(M)+1)$ is a linear combination of compositions of the covariant derivative and the Lie bracket. All relations between these compositions follow from the anticommutativity and the Jacobi identity of the Lie bracket.

The central object will be the subcomplex $\mathcal{Gr}^*_{\bullet\Delta 1}(d)$ of the graph complex $\mathcal{Gr}^*_{\mathrm{Con} \times T^d, T}$ describing 'connected'd-multilinear operators. Its degree m piece $\mathcal{Gr}^*_{\bullet\Delta 1}(d)$ is spanned by connected graphs with dvertices (17) labelled by $X_1, \ldots, X_d$, some number of vertices (18) labelled ∇, m white vertices (19) and one vertex ■. It is clear that $\mathcal{Gr}^*_{\bullet\Delta 1}(d)$ is precisely the subcomplex spanned by connected graphs, of the direct sum $\mathcal{Gr}^*_{\bullet\Delta 1}(d) := \oplus_{c \geq 0} \mathcal{Gr}^*_{\bullet(d)\Delta(c)}$, where $\mathcal{Gr}^*_{\bullet(d)\Delta(c)}$ is the graph complex of [11, Corollary 5.1]. As inProposition 5.4, one easily sees that the collection $\mathcal{Gr}^0_{\bullet\Delta 1} = \{\mathcal{Gr}^0_{\bullet\Delta 1}(d)\}_{d \geq 1}$ forms an operad. It is also not difficult to verify that each graph spanning $\mathcal{Gr}^m_{\bullet\Delta 1}(d)$ has at most 2d+m−1 edges, which explains the stability condition in Theorem 7.2.

Let P={P(d)}$_{d \geq 1}$ be the operad describing algebras with two independent operations—a bilinear product $\star$ satisfying no other conditions and a Lie bracket. Of course, P is the free product (= the coproduct in the category of operads, see [13, p. 137]) of the free operad $\Gamma(\star)$ generated by the bilinear operation $\star$ and the operad Lie for Lie algebras, P=$\Gamma(\star)*\mathcal{L}$ie. Recall that we denoted by $\beta \in \mathcal{L}$ie (2) the generator.

Define the operad homomorphism $F : P \to \mathcal{Gr}^0_{\bullet\Delta 1}$ by F(β):=**b** and F($\star$):=**c**, where $b \in \mathcal{Gr}^0_{\bullet\Delta 1}(2)$ is the graph (39) representing the Lie bracket and $c \in \mathcal{Gr}^0_{\bullet\Delta 1}(2)$ the graph (27) for the covariant derivative. As in Section 5 we easily see that F is well-defined and that $\mathrm{Im}(F) \subset \mathrm{Ker}(\delta : \mathcal{Gr}^0_{\bullet\Delta 1} \to \mathcal{Gr}^0_{\bullet\Delta 1})$. Theorem 7.2clearly follows from

Proposition 7.4

The map $F: P \to \mathcal{Gr}^0_{\bullet \Delta 1}$ induces an isomorphism $P \cong H^0\left(\mathcal{Gr}^0_{\bullet \Delta 1}, \delta\right)$. The generating function $p(t) := \sum_{d \geq 1} \frac{1}{2!} \dim(P(d)).t^d$ for the operad Psatisfies (47)

Proof: The map F embeds into the following diagram of operads and their homomorphisms:

$$P = \Gamma(\star) * \mathcal{Lie} \xrightarrow{\;F\;} \mathcal{Gr}^0_{\bullet \nabla 1} \xrightarrow{\;A\;} \Gamma(\star) * p\mathcal{Lie}$$

$$\downarrow{\scriptstyle T}$$

$$\xrightarrow[\;id * \iota\;]{} \Gamma(\star) * p\mathcal{Lie}$$

$$(48)$$

Let us define the remaining maps in (48). As in [1], one can show that the operad $\mathcal{Gr}^0_{\bullet \Delta 1}$ is isomorphic to the operad $\Gamma(\star) * p\mathcal{L}ie$ governing structures consisting of a bilinear multiplication $\star$ and an independent pre-Lie product $\circ$. The map $\mathcal{Gr}^0_{\bullet \Delta 1} \to \Gamma(\star) * p\mathcal{L}ie$ in (48) is the isomorphism that sends the graph

$\in \mathcal{Gr}^0_{\bullet \nabla 1}(2)$

into $X \star Y \in \Gamma(\star)(2)$ and the graph

$X \in \mathcal{Gr}^0_{\bullet \nabla 1}(2)$

into $X \circ Y \in p\mathcal{L}ie(2)$. The map $T : \Gamma(\star) * p\mathcal{L}ie \to \Gamma(\star) * p\mathcal{L}ie$ is the 'twist' $T(X \star Y) := X \star Y - Y \circ X$ and $T(X \circ Y) := X \circ Y$. It is evident that the composition TAF coincides with the coproduct $id * \iota$ of the identity $id : \Gamma(\star) \to \Gamma(\star)$ and the map $\iota : \mathcal{L}ie \to p\mathcal{L}ie$ given by the antisymmetrization of the pre-Lie product

$\iota([X,Y]):=Y\circ X-X\circ Y$, which is an inclusion by [14, Proposition 3.1]. This implies that $id*\iota$ is a monomorphism, therefore F is a monomorphism, too.

Now, to prove that F induces an isomorphism $\mathcal{P}\cong\mathcal{H}^*\left(\mathcal{G}r_{\bullet\Delta 1}^0,\delta\right)$, it suffices to show that the dimensions of the spaces $\mathcal{H}^0\left(\mathcal{G}r_{\bullet\Delta 1}^0(d),\delta\right)$ and P(d) are the same, for each d1. Our calculation of the dimension of $\mathcal{H}^0\left(\mathcal{G}r_{\bullet\Delta 1}^0(d),\delta\right)$ will be based on the fact that $\left(\mathcal{G}r_{\bullet\Delta 1}^*(d),\delta\right)$ forms a bicomplex. For integers p,q denote by $\mathcal{G}r_{\bullet\Delta 1}^{p,q}(d)$ the subspace of $\mathcal{G}r_{\bullet\Delta 1}^*(d)$ spanned by graphs with precisely $-p$ ∇-vertices. It immediately follows from the replacement rules (29), (30) and (31) that $\delta=\delta'+\delta''$, where $\delta'\left(\mathcal{G}r_{\bullet\Delta 1}^{p,q}(d)\right)\subset\mathcal{G}r_{\bullet\Delta 1}^{p+1,q}(d)$ and $\delta''\left(\mathcal{G}r_{\bullet\Delta 1}^{p,q}(d)\right)\subset\mathcal{G}r_{\bullet\Delta 1}^{p+1,q}(d)$. It is also clear from simple graph combinatorics that the bicomplex $\left(\mathcal{G}r_{\bullet\Delta 1}^{**}(d),\delta\right)$ is bounded by the triangle p=0, p+q=0 and q=d−1, see Fig. 3. The horizontal differential δ' in $\mathcal{G}r_{\bullet\nabla 1}(d)$ is easy to describe—it replaces ∇-vertices according the rule

$$v\ \text{inputs} \qquad\qquad v+2 \qquad\qquad\qquad (49)$$

and leaves other vertices unchanged.

$$\mathcal{G}r_{\bullet\nabla 1}^{-2,2}(3)\xrightarrow{\ \delta'\ }\mathcal{G}r_{\bullet\nabla 1}^{-1,2}(3)\xrightarrow{\ \delta'\ }\mathcal{G}r_{\bullet\nabla 1}^{0,2}(3)$$
$$\uparrow\delta''\qquad\qquad\qquad\uparrow\delta''$$
$$\mathcal{G}r_{\bullet\nabla 1}^{-1,1}(3)\xrightarrow{\ \delta'\ }\mathcal{G}r_{\bullet\nabla 1}^{0,1}(3)$$
$$\uparrow\delta''$$
$$\mathcal{G}r_{\bullet\nabla 1}^{0,0}(3)$$

Figure 3: the bicomplex $\left(\mathcal{G}r_{\bullet\nabla 1}^{**}(3),\delta'+\delta''\right)$.

Remark 7.5

At this point we need to make a digression and observe that $\left(\mathcal{G}r^*_{\bullet\Delta 1}(d),\delta'\right)$ is a particular case of the following construction. For each collection $(U*,\vartheta_U)=\{(U*(s),\vartheta_U)\}s2$ of right dg-Σ_s-modules$(U*(s),\vartheta_U)$, one may consider the complex $\mathcal{G}r^*_{\bullet\Delta 1}\left[U(d)\right]=\mathcal{G}r^*_{\bullet 1}\left[U^*\right](d),\mathcal{V}$ spanned by connected graphs with d vertices (17) labelled $X_1,...,X_d$, one vertex ▮ and a finite number of vertices decorated by elements of U. The grading of $\mathcal{G}r^*_{\bullet 1}\left[U^*\right](d)$ is induced by the grading of U^* and the differential ϑ replaces U -decorated vertices, one at a time, by their ϑ_U-images and leaves other vertices unchanged. It is a standard fact [17] (see also [13, Theorem 21]) that the assignment $\left(U^*,\mathcal{v}_U\right)\to\mathcal{G}r^*_{\bullet\Delta 1}\left[U^*\right](d),v)$ is a polynomial, hence exact, functor, so

$$H^*\left(\mathcal{G}r^*_{\bullet 1}[U^*](d),\vartheta\right)\cong\mathcal{G}r^*_{\bullet 1}\left[H^*(U,\vartheta_U)\right](d)$$

(50)

Let now $(E^*,\vartheta_E)=\{(E^*(s),\vartheta_E)\}_{s\geq 2}$ be such that $E^0(s)$ is spanned by symbols (18), with $v+2=s$,$E^1(s)$ by symbols (19) with $u=s$, and $E^m(s)=0$ for $m\geq$ 2. The differential ϑ_E is defined by replacement rule (49). More formally, $E^0(s)=\mathrm{Ind}\sum^s_{s-2}(1_{s-2})$ and $E^1(s)=1_{s'}$ where 1_s-2 (resp. s1) denotes the trivial representation of the symmetric group Σ_s-2 (resp. Σ_s). The differential ϑE then sends the generator$1\in 1_s-2$ into $-1\in 1_s$. It is clear that, with this particular choice of the collection $(E*,\vartheta_E)$,

$$\left(\mathcal{G}r^*_{\bullet\nabla 1}(d),\delta'\right)\cong\left(\mathcal{G}r^*_{\bullet 1}[E^*](d),\vartheta\right)$$

.

(51)

Let us continue with the proof of Proposition 7.4. Eqs. (50) and (51) in Remark 7.5 imply that

$$H^*\left(\mathcal{G}r^*_{\bullet\nabla 1}(d),\delta'\right)=\mathcal{G}r^*_{\bullet 1}\left[H^*(E,\vartheta_E)\right](d)$$

.

(52)

Since$\vartheta_E:E^0(s)\to E^1(s)$isanepimorphism,thecollection$H^*(E,\vartheta_E)=\{H^*(E(s),\vartheta_E)\}$ $_{s\geq 2}$ is concentrated in degree 0 and $H^0(E(s),\vartheta_E)$ is the kernel of the map $\vartheta_E:E^0(s)\to E^1(s)$. We conclude that $\mathcal{G}r^*_{\bullet 1}\left[H^*(E,\mathcal{v}_E)\right](d)$ is spanned by graphs with d vertices (17) labelled $X_1,...,X_d$, one vertex ▮ and some number of vertices decorated by the collection $H^0(E,\vartheta_E)=\{H^0(E(s),\vartheta_E)\}s2$.

In particular, the graded space $\mathcal{G}r_{\bullet 1}^*\big[H^*(E, v_E)\big](d)$ and hence, by (52), also the horizontal cohomology $H^*\big(\mathcal{G}r_{\bullet \Delta 1}^*, \delta\big), d_1\big)$, is concentrated in degree 0. This implies that the first term $\big(E_1^{p,q}, d_1\big) = \big(H^p\big)\big(\mathcal{G}r_{\bullet \Delta 1}^{*,q}, \delta\big), d_1\big)$ of the corresponding spectral sequence is supported by the diagonal $p+q=0$, so this spectral sequence degenerates at this level and

$$\dim\big(H^0(\mathcal{G}r_{\bullet \nabla 1}^*(d), \delta)\big) = \dim\big(H^0(\mathcal{G}r_{\bullet \nabla 1}^*(d), \delta')\big) = \dim\big(\mathcal{G}r_{\bullet 1}^0\big[H^0(E, \vartheta_E)\big](d)\big).$$

Denote the common value of the dimensions in the above display g_d. We claim that the sequence $\{g_d\}_{d \geq 1}$ satisfies the recursion:

$$\frac{g_{n+1}}{(n+1)!} = \frac{g_n}{n!} + \frac{1}{2!}\sum_{i+j=n}\frac{g_i g_j}{i!\,j!} + \frac{1}{3!}\sum_{i+j+k=n}\frac{g_i g_j g_k}{i!\,j!\,k!} + \frac{1}{4!}\sum_{i+j+k+l=n}\frac{g_i g_j g_k g_l}{i!\,j!\,k!\,l!} + \cdots$$

$$+ \frac{2(2-1)-1}{2!}\sum_{i+j=n+1}\frac{g_i g_j}{i!\,j!} + \frac{3(3-1)-1}{3!}\sum_{i+j+k=n+1}\frac{g_i g_j g_k}{i!\,j!\,k!} + \cdots. \tag{53}$$

This can be seen as follows. Graphs G spanning $\mathcal{G}r_{\bullet 1}^0\big[H^0(E, v_E)\big](d)$ are rooted trees with a distinguished vertex (= root) *. The vertex of G adjacent to the root might either be a vertex (17) or a vertex decorated by $H^0(E, \vartheta_E)$. The contribution from trees of the first type is reflected by the first line of (53), in which the coefficients $1, 1/2!, 1/3!, \ldots$ equal $\dim(1s)/s!$, $s \geq 1$, where $1s$ is the trivial representation of the symmetric group Σ_s spanned by the vertex (17) with $u=s$. The second line of (53) counts contributions from trees of the second type. The coefficients are $\dim(H^0(E(s),_{\vartheta E}))/s!$, $s \geq 2$. It is simple to assemble (53) into Eq. (47).

Let us show that the generating function $p(t) := \sum_{d \geq 1}\frac{1}{d}\dim(p(d)).t^d$ for the operad P also satisfies (47). Since P is, as the coproduct of quadratic Koszul operads, itself quadratic Koszul, one has the functional equation [6, Theorem 3.3.2]:

$$q(-p(t)) = -t. \tag{54}$$

relating p with the generating function $q(t) := \sum_{d \geq 1}\frac{1}{d!}\dim(Q(d)).t^d$ of its quadratic dual Q.

For convenience of the reader, we make a digression and briefly recall the definition of quadratic operads and their quadratic duals. Details can be found in [16, II.3.2] or in the original source [6]. An operad A isquadratic if it is the quotient $\Gamma(E)/(R)$ of the free operad $\Gamma(E)$ on the right $\Sigma 2$-module $E:=A(2)$ of arity-two operations of A, modulo the operadic ideal (R) generated by some subspace $R\subset\Gamma(E)(3)$.

Each quadratic operad $A=\Gamma(E)/(R)$ as above has its quadratic dual A! [16, Definition II.3.37] defined as follows. Let us denote $E^V:=E^*\otimes sgn_2$ the linear dual of the right Σ_2-module E twisted by the signum representation. One then has a natural isomorphism $\Gamma(E^V)(3)\cong\Gamma(E)(3)^*$ of right $\Sigma 3$-modules. Let $R\perp\subset\Gamma(E^V)(3)$ denote the annihilator of R in $\Gamma(E^V)(3)\cong\Gamma(E)(3)^*$. The quadratic dual of A is the quotient $A^!:=\Gamma(E^V)/(R^\perp)$.

To describe the quadratic dual Q of the operad P introduced on page 274 is an easy task. The operad Qgoverns algebras V with two bilinear operations, $\bullet$ and $*$, such that $\bullet$ is commutative associative, $*$ is 'nilpotent' $(a*b)*c=a*(b*c)=0$, $a,b,c\in V$, and these two operations annihilate each other:$(a\bullet b)*c=a*(b\bullet c)=a\bullet (b*c)=0$, $a,b,c\in V$. It is immediately obvious that

$$\dim(\mathcal{Q}(1)) = 1, \quad \dim(\mathcal{Q}(2)) = 3 \quad \text{and} \quad \dim(\mathcal{Q}(d)) = \dim(\mathcal{C}om(d)) = 1 \quad \text{for } d \geqslant 3,$$

where $\mathcal{C}$ om denotes the operad for commutative associative algebras. The generating function for Qtherefore equals $q(t)=e^t-1+t^2$ and Eq. (54) gives $e^{-p(t)}-1+p(t)^2=-t$, which is equivalent to(47). We proved that the generating functions g(t) and p(t) satisfy the same functional equation and, by definition, the same initial condition $p(0)=g(0)=0$, therefore they coincide and $\dim\left(H^0\left(\mathcal{G}r^0_{\bullet\Delta 1}(d),\delta\right)\right) = \dim(P(d))$ for each d1.

In the rest of this section we study operators in $\mathfrak{Nat}_1(\text{Con}\times T^{x\infty},\mathfrak{R})$. Roughly speaking, we prove that all operators in this space are traces in the following sense. Let $\mathcal{D}\in\mathfrak{Nat}(\text{Con}\times T^{x\infty},T)$ be an operator acting on vector fields $X_0,X_1,X_2,\ldots$ and a connection Γ. Suppose that $\mathcal{D}$ is a linear order 0 differential operator in X_0. This means that the local formula $O(X_0, X_1, X_2,\ldots, \Gamma)\in\mathbb{R}$ for $\mathcal{D}$ is a linear function of X_0 and does not contain derivatives of X0. For such an operator we defineTrX$_0$ ($\mathcal{D}$) $\in\mathfrak{Nat}(\text{Con}\times T^{x\infty},\mathfrak{R})$ by the local formula

$$Tr_{X_0}(O)(X_1, X_2, \ldots, \Gamma) := \text{Trace}\big(O(-, X_1, X_2, \ldots, \Gamma) : \mathbb{R}^n \to \mathbb{R}^n\big) \in \mathbb{R}.$$

It is easy to see that $TrX_0(\mathfrak{O})$ is well defined. Let us formulate a structure theorem for multilinear operators from $\mathfrak{Nat}_1(\text{Con} \times T^{\times \infty}, \mathfrak{R})$.

Theorem 7.6

Let d0. On smooth manifolds of dimension 2d, each d-multilinear operator in $\mathfrak{Nat}_1(\text{Con} \times T^{\otimes d}, \mathfrak{R})$ is the trace of a $(d+1)$-multilinear operator from $\mathfrak{Nat}_1(\text{Con} \times T^{\otimes(d+1)}, T)$.

Theorem 7.6 will follow from Proposition 7.8 below. A depolarized version of Theorem 7.6 is:

Corollary 7.7

On a smooth manifold M, each operator from $\mathfrak{Nat}_1(\text{Con} \times T^{\times \infty}, \mathfrak{R})$ whose all components are of homogeneity $\leq \frac{1}{2} \dim(M)$ is a trace of an operator from $\mathfrak{Nat}_1(\text{Con} \times T^{\times \infty}, T)$.

Denote by $\mathcal{Gr}^*_{\bullet \Delta \circlearrowleft}(d)$ the graph complex describing operators in $\mathfrak{Nat}_1$ $(\text{Con} \times T^{\otimes d}, \mathfrak{R})$. The degree m-component of this complex is spanned by connected graphs with d vertices (17) labelled $X_1, \ldots, X_d$, some number of vertices (18) labelled ∇ and m white vertices (19). It is not difficult to see that the number of edges of graphs spanning $\mathcal{Gr}^*_{\bullet \Delta \circlearrowleft}(d)$ is 2d, which explains the stability assumption in Theorem 7.6.

We will also consider the subcomplex $\mathcal{Gr}^*_{\bullet \Delta Tr}(d) \subset \mathcal{Gr}^*_{\bullet \Delta 1}(d+1)$ of graphs describing operators in $\mathfrak{Nat}_1(\text{Con} \times T^{\otimes(d+1)}, T)$ for which the trace is defined. Clearly, the degree m component $\mathcal{Gr}^m_{\bullet \Delta Tr}(d)$ of this subcomplex is spanned by connected graphs with one vertex $\ast$ labelled X_0, one vertex $\dagger$, d vertices(17) labelled $X_1, \ldots, X_d$, a finite number of vertices (18) labelled ∇ and m white vertices (19). The trace is represented by the map $Tr : \mathcal{Gr}^*_{\bullet 1 \nabla Tr}(d) \to \mathcal{Gr}^*_{\bullet \nabla \circlearrowleft}(d)$ that removes the vertices $\ast$ and $\dagger$ and connects the two loose edges created in this way by a directed wheel. It is clear that this map commutes with the differentials. We now establish Theorem 7.6 by proving the following.

Proposition 7.8

The map $\mathrm{Tr}: \mathcal{Gr}^*_{\bullet\nabla\mathrm{Tr}}(d), \delta) \to \mathcal{Gr}^*_{\bullet\nabla\circlearrowleft}(d), \delta)$ induces an epimorphism of cohomology $H^0\left(\mathcal{Gr}^*_{\bullet\nabla\mathrm{Tr}}(d), \delta\right) \to H^0\left(\mathcal{Gr}^*_{\bullet\nabla\circlearrowleft}(d), \delta\right)$.

Proof: As in the proof of Proposition 7.4 we observe that both $\left(\mathcal{Gr}^*_{\bullet\nabla\mathrm{Tr}}(d), \delta\right)$ and $\left(\mathcal{Gr}^*_{\bullet\nabla\circlearrowleft}(d), \delta\right)$ are bicomplexes, with $\mathcal{Gr}^{p,q}_{\bullet\nabla\circlearrowleft}(d)$ (resp. $\left(\mathcal{Gr}^{p,q}_{\bullet\nabla\circlearrowleft}(d)\right)$) spanned by graphs in $\mathcal{Gr}^{p,q}_{\bullet\nabla\mathrm{Tr}}(d)$ (resp. $\left(\mathcal{Gr}^{p,q}_{\bullet\nabla\circlearrowleft}(d)\right)$) with precisely $-p$ ∇-vertices. The differential in both complexes decomposes as $\delta = \delta' + \delta''$ where δ' (the 'horizontal part') raises the p-degree by one and preserves the q -degree, and δ'' (the 'vertical part') preserves the q-degree and raises the p-degree by one.

The map $\mathrm{Tr}: \mathcal{Gr}^*_{\bullet\nabla\mathrm{Tr}}(d), \delta) \to \mathcal{Gr}^*_{\bullet\nabla\circlearrowleft}(d), \delta)$ obviously preserves the bigradings, therefore it induces the map

$$H^*(\mathrm{Tr}, \delta') : H^*\left(\mathcal{Gr}^*_{\bullet\nabla\mathit{Tr}}(d), \delta'\right) \to H^*\left(\mathcal{Gr}^*_{\bullet\nabla\circlearrowleft}(d), \delta'\right) \tag{55}$$

of the horizontal cohomology. Using the same considerations as in the proof of Proposition 7.4, we identify this map with

$$\mathit{Tr} : \mathcal{Gr}^*_{\bullet\mathit{Tr}}\left[H^*(E, \vartheta_E)\right](d) \to \mathcal{Gr}^*_{\bullet\circlearrowleft}\left[H^*(E, \vartheta_E)\right](d) \tag{56}$$

where (E^*, ϑ_E) is the dg-collection introduced in Remark 7.5 and the graph complexes in (56) are defined analogously as the graph complex $\mathcal{Gr}^*_{\bullet\mathbin{|}}\left[H^*(E, \vartheta_E)\right](d)$ used in the proof of Proposition 7.4.

Let us show that the map in (56) is an epimorphism. Consider a graph G in $\mathcal{Gr}^*_{\bullet\circlearrowleft}\left[H^*(E, \vartheta_E)\right](d)$ and choose a directed edge e in the (unique) wheel of G. Let $\bar{G}$ be the graph in $\mathcal{Gr}^*_{\bullet\mathit{Tr}}\left[H^*(E, \vartheta_E)\right](d)$ obtained by cutting e in the middle and decorating the loose ends thus created by vertices ♠ and ♯ as in the following display:

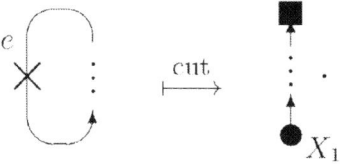

Clearly $\mathrm{Tr}(\hat{G}) = G$ which proves that (56) is surjective. So, we have two spectral sequences, $(E^{p,q}_*, d_*)$ and $(F^{p,q}_*, d_*)$, such that

$$(E^{p,q}_0, d_0) = (\mathfrak{Gr}^{p,q}_{\bullet\nabla Tr}(d), \delta), \qquad (F^{p,q}_0, d_0) = (\mathfrak{Gr}^{p,q}_{\bullet\nabla\circlearrowright}(d), \delta)$$

and the map $\mathrm{Tr}_* : (E^{p,q}_*, d_*) \to (F^{p,q}_*, d_*)$ induced by the trace map $\mathrm{Tr}_* : \mathfrak{Gr}^*_{\bullet\nabla Tr}(d) \to \mathfrak{Gr}^*_{\bullet\nabla\circlearrowright}(d)$. The map $\mathrm{Tr}_1 : (E^{p,q}_1, d_1) \to (F^{p,q}_1, d_1)$ of the first levels of the spectral sequences is (55) and we identified this map with epimorphism (56). It is also clear that the first terms of both spectral sequences are supported by the diagonal p+q=0, so these spectral sequences degenerate at this level. A standard argument then implies that the map $H^0(\mathrm{Tr}) : H^0(\mathfrak{Gr}^*_{\bullet\nabla Tr}(d), \delta) \to (\mathfrak{Gr}^*_{\bullet\nabla\circlearrowright}(d), \delta)$ in Proposition 7.8 is an epimorphism.

ACKNOWLEDGMENTS

I would like to express my thanks to S. Merkulov for sharing his ideas with me, to A. Alekseev for suggesting an interpretation of the homological vector field in terms of the Chevalley–Eilenberg differential, and to G. Weingart who pointed some flaws in my reasoning to me. Also conversations with J. Janyška and J. Slovák were extremely useful. Suggestions of the referee lead to a substantial improvement of the paper.

REFERENCES

1. F. Chapoton, M. Livernet, Pre-Lie algebras and the rooted trees operad, Internat. Math. Res. Notices 8 (2001) 395–408.
2. J. Conant, Fusion and fission in graph complexes, Pacific J. Math. 209 (2) (2003) 219–230.
3. J. Conant, K. Vogtmann, Infinitesimal operations on complexes of graphs, Math. Ann. 327 (3) (November 2003) 545–573.
4. J. Conant, K. Vogtmann, on a theorem of Kontsevich, Algebr. Geom. Topol. 3 (2003) 1167–1224.
5. A.S. Dzhumadil'daev, 10-commutator and 13-commutator, Preprint math-ph/0603054, March 2006.
6. V. Ginzburg, M.M. Kapranov, Koszul duality for operads, Duke Math. J. 76 (1) (1994) 203–272.
7. P.I. Katsylo, D.A. Timashev, Natural differential operations on manifolds: an algebraic approach, Preprint math.DG/0607074, July 2006.
8. I. Kolár, P.W. Michor, J. Slovák, Natural Operations in Differential Geometry, Springer-Verlag, Berlin, 1993. ˇ
9. M. Kontsevich, Formal (none) commutative symplectic geometry, in: The Gel'fand Mathematics Seminars 1990–1992, Birkhäuser, 1993.
10. D. Krupka, Elementary theory of differential invariants, Arch. Math. (Brno) XIV (1978) 207–214.
11. M. Markl, Invariant tensors and graphs, Preprint arXiv: 0801.0418, available from www.arXiv.org.
12. M. Markl, Homotopy algebras via resolutions of operads, in: Proceedings of the 19th Winter School "Geometry and physics", Srní, Czech Republic, January 9–15, 1999, Supplem. ai Rend. Circ. Matem. Palermo, Ser. II, vol. 63, 2000, pp. 157–164.
13. M. Markl, Homotopy algebras are homotopy algebras, Forum Math. 16 (1) (January 2004) 129–160.
14. M. Markl, Lie elements in pre-Lie algebras, trees and cohomology operations, J. Lie Theory 17 (2) (2007) 241–261.
15. M. Markl, S.A. Merkulov, S. Shadrin, Wheeled PROPs, graph complexes and the master equation, Preprint math.AG/0610683, October 2006.
16. M. Markl, S. Shnider, J.D. Stasheff, Operads in Algebra, Topology and Physics, Mathematical Surveys and Monographs, vol. 96, American Mathematical Society, Providence, Rhode Island, 2002.
17. M. Markl, A.A. Voronov, PROPped up graph cohomology, Preprint math. QA/0307081, July 2003.
18. S.A. Merkulov, Operads, deformation theory and F -manifolds, in: Frobenius Manifolds, in: Aspects Math., vol. E36, Vieweg, Wiesbaden, 2004, pp. 213– 251.
19. S.A. Merkulov, PROP profile of deformation quantization, Preprint math. QA/0412257, December 2004.

20. S.A. Merkulov, Graph complexes with loops and wheels, Preprint, Duke Math. J., 2006, submitted for publication.
21. S.A. Merkulov, PROP profile of Poisson geometry, Comm. Math. Phys. 262 (1) (February 2006) 117–135.
22. M. Mulase, M. Penkava, Ribbon graphs, quadratic differentials on Riemann surfaces, and algebraic curves defined over Q, Asian J. Math. 2 (4) (1998) 875–919. Mikio Sato: a great Japanese mathematician of the twentieth century.
23. R.S. Palais, C.L. Terng, Natural bundles have finite order, Topology 19 (3) (1977) 271–277.
24. M. Penkava, Infinity algebras and the cohomology of graph complexes, Preprint q-alg/9601018, January 1996.
25. M. Penkava, A. Schwarz, A∞ algebras and the cohomology of moduli spaces, Trans. Amer. Math. Soc. 169 (1995) 91–107.
26. C.L. Terng, Natural vector bundles and natural differential operators, Amer. J. Math. 100 (1978) 775–828.
27. G. Weingart, Local covariants in Cartan geometry, Preprint, November 2006.

CITATION

1. Martin Markl, Natural differential operators and graph complexes, Differential Geometry and its Applications, Volume 27, Issue 2, April 2009, Pages 257-278, ISSN 0926-2245, doi. 10.1016/j.difgeo.2008.10.008.

Index